A Rainbow Palate

synthesis

Synthesis

A series in the history of chemistry, broadly construed, edited by Carin Berkowitz, Angela N. H. Creager, John E. Lesch, Lawrence M. Principe, Alan Rocke, and E. C. Spary, in partnership with the Science History Institute

A Rainbow Palate
How Chemical Dyes Changed the West's Relationship
with Food

Carolyn Cobbold

The University of Chicago Press :: Chicago and London

The University of Chicago Press, Chicago 60637
The University of Chicago Press, Ltd., London
© 2020 by The University of Chicago
Published 2020
Printed in the United States of America

29 28 27 26 25 24 23 22 21 20   1 2 3 4 5

ISBN-13: 978-0-226-72705-9 (cloth)
ISBN-13: 978-0-226-72719-6 (e-book)
DOI: https://doi.org/10.7208/chicago/9780226727196.001.0001

Library of Congress Cataloging-in-Publication Data

Names: Cobbold, Carolyn, author.
Title: A rainbow palate : how chemical dyes changed the West's
    relationship with food / Carolyn Cobbold.
Other titles: Synthesis (University of Chicago. Press)
Description: Chicago : University of Chicago Press, 2020. | Series:
    Synthesis | Includes bibliographical references and index.
Identifiers: LCCN 2020004022 | ISBN 9780226727059 (cloth) |
    ISBN 9780226727196 (ebook)
Subjects: LCSH: Coloring matter in food. | Adulterations.
Classification: LCC TP456.C65 C633 2020 | DDC 664/.062—dc23
LC record available at https://lccn.loc.gov/2020004022

♾ This paper meets the requirements of ANSI/NISO Z39.48-1992
(Permanence of Paper).

Contents

Preface

Aniline and azo dyes were the first of many novel substances including drugs, perfumes, sweeteners, and flavorings, which chemists began to synthesize and produce on an industrial scale from coal tar, a waste product of the gas industry. Although intended for use in the textile industry, the new dyes soon began to be added to food, becoming one of the first examples of laboratory-created, industrially manufactured chemicals to be used in our daily life in unexpected ways. Chemists and physicians, politicians and campaigners, food and dye producers, retailers and the public jostled to determine and arbitrate the use of these new substances. The new dyes were among the earliest contested chemical additives in food, and the battles surrounding their use offer striking insights and parallels into today's international struggles surrounding chemical, food, and trade regulation.

Chemists maneuvered themselves to become key players in new arrangements for assessing food quality amid cultural differences across Europe and in America that led to different regimes of defining and regulating food products. Strategies for testing these novel ingredients were devised by chemists from a variety of backgrounds and disciplines and with different aims and objectives, drawing on a diverse range of analytical repertoires. Despite being brilliantly colored and increasingly ubiquitous, these new substances remained almost impossible to detect for decades. Although the dyes were created by chemists, the detection and evaluation of aniline dyes in food represented a failure in analytical chemistry. While commerce, politicians, and the public all invoked chemists to represent their interests, the authority, technical ability, and impartiality of the scientists were never sufficient to successfully arbitrate on the issue, leaving the situation still unresolved more than a century later.

We now live in a world saturated with synthetic chemicals, and while our society couldn't function without the hundreds of thousands of synthesized chemicals we use in everything from the food we eat to the packaging surrounding it, our cleaning products and pesiticides, our clothes, cars, and household furnishings, concerns are increasing about the impact they are having on the environment and our health, with our bodies now playing host to hundreds of synthetic chemicals that did not exist before the 1850s. Aniline and azo dyes were the substances that led to the synthethized chemical industry that has reshaped our world and the food we eat.

Introduction

When Judy Garland's Dorothy opened the door of a tornado-wrecked house to discover she had left the black and white world of Kansas and arrived in the rainbow-colored Land of Oz, an indelible imprint was formed on the collective memory of the twentieth century. Her entrance to a land of brilliant reds, yellows, and greens spectacularly highlighted the new technique of Technicolor, a chemical dye–based technology that literally changed moviegoers' view of their world.[1] But there is another scene from 1939's iconic *The Wizard of Oz* that demonstrates a part of the chemical dye revolution that is rarely celebrated, or written about historically. And that, as they say, is a horse of a different color and the subject of this book—how a palette of chemical dyes changed our relationship with food.

How technicians changed the color of the horse pulling the coach in the Emerald City from yellow to red to purple is an enduring Hollywood tale. After film technicians were unable to achieve the horse's color transformation using Technicolor and the American Society for the Prevention of Cruelty to Animals forbade the dyeing of the horse, the film's producers turned to Jell-O, one of America's favorite chemically dyed food products. Look very carefully and you can even see the horse trying to lick the lemon-, cherry-, and grape-flavored jelly smeared all over it.[2]

The chemical dyes that helped create Jell-O and other industrially processed foods that transformed the Western diet were essentially the same dyes that produced the Technicolor prints and rainbow costumes of *The Wizard of Oz*.

As described by Anthony Travis in *The Rainbow Makers: The Origins of the Synthetic Dye Industry in Western Europe*, the aniline and azo dyes had been discovered and synthesized by chemists from coal tar starting in the late 1850s.[3] By 1900 the bright coal tar dyes had transformed the world of the consumer. Paintings, textiles, walls, crockery, clothes, glass, and food all shone brighter.[4] Novelists and newspapers spun tales about the chemistry and colors that were transmogrifying life in Europe and America. Authors from Charles Dickens to L. Frank Baum described a world where chemists created purple ribbons and magic colored bonbons, not to mention a fantastical land featuring a yellow brick road and an Emerald City.[5] Baum was acutely aware of how the bright aniline and azo dyes changed consumer goods, particularly the fashion industry. Prior to writing *The Wonderful Wizard of Oz*, Baum edited *The Show Window* and was a founder of the National Association of Window Trimmers of America.[6] The textiles and clothes that adorned shop windows in the last few decades of the nineteenth century were public showcases for the brilliant new dyes.

The dyes were the first of many new chemical substances, including drugs, fertilizers, perfumes, sweeteners, and flavorings, that chemists began to synthesize and produce on an industrial scale from coal tar, a waste product of the gas industry. Chemists eagerly sought to boost their professional status by riding the wave of new brilliant colors transforming the Western world, as the press attributed the wondrous dyes to the science of chemistry and the skills of chemists. However, chemists struggled to retain control of their creations once they left the laboratories and factories and began to be used in novel and unintended ways in the marketplace. Chemists created the new colors specifically for the textile industry, but it was the marketplace participants, including manufacturers, retailers, consumers, politicians, and the press, who determined how the dyes were used. As a result, the dyes were among the first examples of laboratory-created, industrially manufactured chemicals to permeate our daily life, in unintended and unpredictable ways.

When concerns surrounding the possible toxicity of the dyes being used in food began to surface, chemists sought to influence how the new substances were used and perceived in the marketplace. In many ways, the reputation of chemists and chemistry were irrevocably linked to the dyes. Chemists argued that they were best placed to determine the safety and appropriate use of chemical dyes in food, but they were just one group among many, including food manufacturers, the press, the public, and politicians, who sought to determine the policing of the new dyes and how and where they could be used. By the time risks and uncertainties surrounding the synthesized chemicals began to surface, the dyes were being used everywhere from clothes and home furnishings to crockery and food.

The political, public, and chemical contestations surrounding the transmutation of black and grimy industrial waste into the brilliantly colored clothes and food of the late nineteenth century are worthy of any tale by Baum or Dickens. The history of how aniline and azo dyes became the first man-made, industrially produced chemicals to become legalized food additives chimes with contemporary debates about "legal adulteration."[7] It is also particularly timely as today's public and media increasingly pour scorn on the twentieth-century promises of "better living through chemistry" and question the safety of the hundreds of thousands of synthetic chemicals developed since the first chemical dyes were manufactured.

This book focuses on Britain, as one of the first countries to manufacture aniline dyes commercially and to introduce legislation to regulate the sale of adulterated food but one of the last major Western nations to introduce prescriptive legislation banning or permitting food colorings. However, unlike previous histories of either the food or the dye industries, this book is a horse of a different color, galloping from Britain and France to Germany and the US, tracing the network of chemists working in the chemical manufacturing and food production industries and in public analysis of food. By describing the conflicting roles of the various chemists involved in the production, use, and control of dyes, it is also a horse that cannot be confined to a single stable. The history of aniline and azo dyes, synthesized from coal tar and consumed in unforeseen and contentious ways, is a history of chemistry, of food, of color and consumer culture, of economics, risk management, and regulation. Imbued with politics and pragmatism and disputed boundaries, it demonstrates how contested the control of new science becomes once it emerges from the laboratory and enters the marketplace.

Until recently, few historians had examined the use of chemical dyes in food.[8] This oversight was perplexing when one considers the extensive public debate about food colorings that has been ongoing since the 1950s and heightened in the 1980s and continues today. Several of the original coal-tar derivative food colorings, now innocuously named red 1, blue 2, or yellow 5, have been linked with cancer or behavioral problems, yet synthetic food dyes are still widely used around the world.[9] Some countries began to ban such dyestuffs in the late 1800s. Many countries now prohibit them totally, while other nations have identified a limited number of "safe" dyes. However, some countries, particularly developing countries, have few restrictions on synthetic food dyes. At the same time, multinational food manufacturers choose to market some of their products as being free from artificial coloring, yet are prepared to use such additives in other products. As a result, more than 150 years after synthetic dyes were introduced into our food, the debate over their benefits and risks remains unresolved.

When William Perkin accidentally discovered a new purple dye he called mauveine while experimenting with coal tar in 1856, he immediately saw its potential outside the laboratory as a textile dye.[10] Chemists in the decades to follow produced an amazing new palette of chemical dyes from combining hydrocarbons such as benzene and toluene and xylene obtained from the distillation of coal with other compounds. Perkin's "metamorphosis" of matter, from coal tar into a new dye that had not existed before, led to transformations of matter on an industrial scale and created a new commercial value attached to science.[11]

How these new chemical substances became consumed as food additives amid increasing concerns surrounding food quality is a topic that resonates in today's climate of distrust in food provenance and technology.[12] Food is a particularly useful commodity for studying complex relations between makers, users, legislators, and scientific practitioners because it is universal. All members of a society must ingest food into their bodies. The fusion of such a fundamental commodity as food with totally new chemical substances, synthetically produced from what was essentially a waste commodity, links in fascinating ways to debates surrounding taste, trust, and truths.[13]

Of course, issues of food purity and safety and debates about food coloring and additives are much older than the introduction of chemical dyes, and the role of chemists in the food system also predates the nineteenth century. Equally, consumer and public health concerns surrounding the use of aniline and azo dyes in food continued to be debated internationally throughout the twentieth century and are still of a matter of concern and debate. Indeed, among historians there is increasing interest in the role of chemical additives, including dyes, in twentieth-century food, as evidenced by Angela Creager's "Risk on the Table: Food, Health, and Environmental Exposures" project.

However, the late nineteenth century was a game-changing period that ultimately led to the chemogastric revolution and our synthetic chemical world. This was a time when food production was becoming increasingly industrialized, with consumers facing complex and contradictory food knowledge claims. Industrial chemists and food manufacturers were seeking to introduce synthetic colorings during a period when food adulteration was of considerable social concern. At the same time, analytical chemists were being paid to identify harmful and fraudulently applied food additives, raising questions of whom to trust and how "scientific" knowledge is formed and evaluated.[14]

Chemists transformed both dark and dirty industrial waste and natural matter into artificial substances that could replicate the qualities of nature itself. In a seemingly alchemical act of transmutation, they synthesized the molecules of coal tar from dead, dark matter that had laid for centuries in the depths of the Earth into new substances that would transfigure society and science. However, while hailed initially as the "Wonders of Coal-Tar," these new substances soon began to be perceived by some as dangerous and unnatural. The transmutation from a black pollutant into the source of medicinal drugs and wondrous colors, as well as substances of unknown toxicity, may be seen as a metaphor for the transformations of science and scientific practitioners during the same period.[15] The significance and meaning attached to these transformations can be more deeply appreciated by examining the reactions of different individuals and institutions when the new substances were incorporated into food, to be ingested by everyone.

The Victorian era has been variously described by historians and nineteenth-century contemporaries as an "Age of Transition," with pronounced transformations in bureaucracy, government, customs and habits, science and technology, politics and commerce; an "Age of Progress and Optimism," with an underlying belief in rational science, free trade, prosperity, and democracy; but also an "Age of Anxiety and Degeneration," with customs and beliefs, religious and moral order under threat from scientific rationalism, urbanization and industrialization.[16] All these cultural changes and attitudes affected how the new chemical substances were regarded and consumed.

Coal tar dyes were introduced into food across Europe and the US at a time when these countries were busy constructing and amending food legislation to combat adulteration. As some historians are now recognizing, and which will become increasingly apparent in this history, regulation also effectively legitimized some forms of food adulteration, including the use of new synthetic chemicals as ingredients.[17] However, each country adopted a different strategy for addressing coal tar dyes. France and Germany banned specified coal tar dyes known to be harmful from the 1880s, while the US, in 1907, advised against the use of all coal tar dyes except for seven permitted dyes. Britain, meanwhile, did not specifically mention any coal tar dyes in legislation before 1925 and only produced a permitted list of chemical dyes in 1957, five decades after the US.[18] Different cultural, political, and institutional frameworks in each country influenced these differing regulatory approaches.

The public transformation of chemical dyes

The burgeoning nineteenth-century media in countries such as Britain, including newspapers and novels as well as consumer, trade, and professional journals, was central to the public transformation of the dyes' meaning. News reporting both reflected and formed public perception, making the information flow a two-way process. From the 1850s onwards, this was an era when advances in technology, including the rotary press and telegraph, as well as a reduction in stamp duty on newspapers, had led to an explosion of newspapers and journals, a time when the newspaper became part of daily life for many people. The reporting of news as it was viewed throughout the twentieth century was created during the nineteenth century, described by Matthew Rubery as an "explicitly commercial commodity—that is, impersonal information sold for the purposes of commerce, communication, or pleasure."[19] Science, food adulteration, public health and hygiene, and government legislation were all issues covered extensively in the nineteenth-century press.[20]

The original profile of the new dyes, determined by scientists, was mediated by a wide range of readers and writers, both lay and scientific, into something with a different presence and significance. The changed perception of the dyes and their use in food led to transformations in the substances themselves and their applications as well as those manipulating them. Manufacturers, retailers, scientists, politicians, and social reformers used the media to further their own interests, and all played a part in the metaphysical transformation of the new dyes. The changing media representation of the dyes and their use in food played an important part in the perception of chemists and chemistry as well as of the dyes themselves.

Harvey Levenstein claims that present-day food fears and concerns about industrialized food can be dated back to late nineteenth- and early twentieth-century debates about food, legislation, and chemical additives. He argues that the "slew of fearsome stories about poisonous food additives actually sowed the seeds for the fears of processed foods which periodically sweep the nation in the years to come."[21] This book will show, however, that attempts by chemists and regulators to mediate the use of synthesized chemicals in food and to reassure the public actually helped to legitimize their use, effectively converting them from adulterants to ingredients.

Industrial production of dyes

The economic significance of the substances extracted from coal tar was immediately recognized by perceptive commentators of the day, such as Charles Dickens who noted how "alchemists of old spent their days and nights searching for gold, and never found the magic Proteus, though they chased him through all gases and all metals. If they had, indeed, we doubt much if the discovery had been as useful as this of Perkins's [sic] purple . . . A discovery that benefits trade is better for a man than finding a gold mine."[22]

The link between chemistry and the dye industry is a long-standing and strong one, with dyeing considered in the eighteenth century as one of the "cultures of chemistry." Agustí Nieto-Galan has shown that the production of "natural" dyes from vegetables such as madder, woad, annatto, and indigo, insects such as cochineal and kermes, and minerals such as lead white and Prussian blue, a ferric iron cyanide, involved considerable chemical knowledge and synthesis. As with food itself, what is "natural" and what is constructed is a relative and contested issue and even the production of "natural" dyes involved considerable chemical knowledge and a degree of synthesis. Prior to the mid-nineteenth century, many chemists were involved in the dyestuff industry, including Louis-Jacques Thénard, Joseph Louis Gay-Lussac, the industrial chemist Jean-Antoine Chaptal, and Michel-Eugène Chevreul, director of dyeing at the Gobelins tapestry factory. Factories making and using chemicals for the dyeing industry grew in size and technical complexity from the mid-eighteenth century onwards, with some calico-printing factories employing upwards of a thousand workers.[23] However, as Andrew Pickering observes, the production of coal tar dyestuffs was on a wholly different scale, with the new synthetic colors manufactured in a laboratory and not extracted from nature.[24] In many ways, these new industrially manufactured objects represented what science was capable of achieving. Thus, their reputation was particularly tied to the status of chemistry and chemists.

For the first half of the nineteenth century, bright colors were associated with the far-flung exotic reaches of empire, such as the indigo dyes of India and the cochineal beetles of the New World. However, the discovery of hundreds of vibrant colors capable of being extracted from coal tar proved to be a pivotal moment, transforming material culture and aesthetic values. Colors became the epitome of science and industry and central to the economy of civilization. Michael Taussig claims that the history of modern coloring is closely aligned with that of colonialism and subsequently global capitalism. In his view, the "brave new world of artifice created by chemical magic was to Germany what empire was to Britain and France."[25] Germany's dominance in coal tar color technology was partly driven by this new nation's attempt to create a techno-scientific empire to rival the colonial-based empires of France and Britain.

By 1880, Germany was by far the world's largest producer of synthetic dyes. Indeed, there is a strong German focus to the rise of synthetic dye production and consumption throughout the industrialized West, for many German-born and German-educated chemists travelled and worked throughout Europe and the US. In so doing, they spread a knowledge and understanding of organic chemistry that was fashioned to a large extent in Germany. Prior to 1850, organic chemists produced and analyzed many synthetic substances.[26] However, the production of chemical dyes from coal tar waste radically altered the economic significance of organic chemistry and led to the formation of huge chemical conglomerates that are still operating in Germany today, including Badische Anilin und Soda Fabrik (BASF), Aktien Gesellschaft für Anilin-Fabrikaten (AGFA), Hoechst, and Bayer.

Industrial production of food

Dyes were not the only commodities whose production was drastically transformed by industrialization. The second half of the nineteenth century was a period when food, including staple items such as bread, milk, and meat, was becoming increasingly industrialized and processed. Many changes in food production, distribution, and consumption occurred, notably a move away from domestic provision and preparation. These changes were partly enabled by the development of preservatives both chemical and physical, such as canning and refrigeration, resulting in less dependency on seasons and the ability to move food over greater distances.[27] Often production and preservation techniques altered the natural coloring of foods, making the use of artificial food colorings more desirable. Fossil fuel powered the steam machines that processed the food and transported it over long distances; its waste product, in the form of coal tar dyes, helped manufacturers, wholesalers, and retailers to restore the colors lost through processing and, in many cases, to increase profits.

By the end of the nineteenth century, food companies in the industrial West, such as Heinz, Rowntree's, and Huntley and Palmer's, were economically and politically powerful. Realizing that regulation would benefit them more than their smaller competitors, large companies actively encouraged and helped shape legislation and used adulteration fears in promoting and advertising their products.[28] Food manufacturers began to use consultant chemists and analysts to deflect concerns surrounding adulteration. They also saw the benefits of chemical and scientific techniques in reducing costs, increasing brand differentiation, and obtaining consistent quality. As a result, large food manufacturers began to employ chemists from the turn of the century onwards, with several having established company laboratories within a few years.[29] Indeed, investigations into dye colors were among the first tasks undertaken by chemists working for food producers.

Lists of ingredients from food company invoices, recipe books, and experimental notebooks show that hundreds of new chemical substances, particularly dyes, entered an increasingly commodified food supply. The new dyes were sold in the form of liquid, paste, or powder and used to color a wide range of food products such as butter, cheese, noodles, wines, confectionery, wines, and liquors. However, partly because of the tiny quantities of dye used, even expert chemists were often unable to detect their use, still less identify the individual dyes. These novel substances, created at the cutting edge of chemical science, came to permeate food and drink products invisibly across the Western world, passing through a circuitous and opaque international supply chain from chemical factory to wholesale and retail chemists to the food industry, with the physical and chemical origins of these fugitive ingredients becoming more and more obscured.

A glance at the internet and newspapers reveals the array of cultural and "scientific" influences that affect our food color choices, from the "discovery" of the latest red or purple "superfood" to the concerns surrounding artificial colorings and their alleged impact on health.[30] Anthropologists and sociologists argue that food color preferences are culturally and socially determined, varying both between and within societies.[31] Different food colors also take on various symbolic meanings, particularly in religious and celebratory traditions around the world.[32] In recent years, the importance of food color in food choice has become the subject of increasing scientific and nutritional research, including the marketing of color-coded diets.[33] Yet using color to improve food presentation and to assess its quality or health benefits is nothing new. Food company archives, newspaper and journal articles, and transcripts of government committee hearings all reveal the importance of food color to the nineteenth-century consumer and producer.

Testing and understanding new substances

Chemists throughout Europe and the US had to devise new experiments in order to understand the dyes that were increasingly being used to color food. The problem at stake for chemists was how to make knowledge from synthetic dyes, a totally new and unknown commodity with no prior tradition or recognition. Tests were sought to address growing public and political concerns about the use and long-term physiological effects of chemical additives, and to inform the introduction or reform of food legislation. The dyes were elusive both in their detection and legitimation, socially and scientifically, and reaching agreement over the accuracy, standardization, and interpretation of tests to evaluate them was not easy. The apparatus of assessment and control itself had to be modified to accommodate these new substances. Chemists from diverse cultures, countries, and disciplines, with different social status, interests, skills, and practices, tried to establish viable tests and methods to detect and assess the new dyes.[34]

By the nineteenth century, aspiring chemists could purchase an analytical laboratory that could be set up at home, in the field, or in a shop, a development that led to an expansion of practical chemistry and education and the beginnings of a chemical profession.[35] John Pickstone describes the nineteenth century as an "analytical age," with chemists "working for government in the enforcement of industrial and public health legislation that they helped promote."[36] However, he and other historians of science have demonstrated competing tensions in analytical chemistry during this period, notably between pure and applied chemistry.[37]

Examining the tests devised and performed by different chemists to identify and assess entirely new, and unknown, chemical substances allows us further to explore these tensions and the lack of consensus in chemistry during this period. The variety of tests devised demonstrates the multitude of ways of knowing, naming, and understanding the synthetic dyes among a heterogenous group of chemists operating in a variety of settings, from artisanal workrooms, industrial workplaces, and pharmacists' and analysts' kitchens to institutional laboratories. Both the public place and image of chemistry underwent significant and problematic transformation during the second half of the century, and synthetic dyes were central to this change. Chemists were, on the one hand, creating new substances, such as drugs and dyes that promised to transform consumers' way of life in industrial factories, while, on the other hand, detecting poisons and pollutants, including chemically synthesized substances, using portable equipment. However, the inability to detect and assess the dyes as they spread into the marketplace compromised the expertise and status of chemists in the public sphere at a time when they were struggling to establish themselves as food and analytical experts. The situation in assessing chemical dyes was particularly complex, for these were novel and undetectable substances, being created by chemists who themselves were unable to agree on their nomenclature or chemical composition. The assessment of synthetic dyes involved uncertain facts, undetectable substances, and a widespread and disparate group of parties seeking agreement as to their identity, form, use, and consequence.[38] Attaining consensus among a fragmented group of individual chemists from diverse backgrounds, with a range of paymasters, different agendas, and varying social and economic circumstances, was an impossible task.

Food adulteration and the rise of the public analysts

The expanding role of professional chemists as food chemists was closely linked to the industrialization of food and increasing concerns surrounding food adulteration. Previously overlooked as commonplace and self-evident, food's significance and ability to shed light on social history was highlighted by the *Annales* school of historians in the 1960s and '70s, and interest has strengthened since. As E. C. Spary points out, "Food might be legitimated as a historical subject by its very universality: something which affects the life of every person, every day, ought to carry enormous weight in a democratizing historiography."[39]

Reactions against increasing industrialization of food in the nineteenth century produced consumer and political movements across Europe and the US that opposed food adulteration and promoted projects for returning to "natural" foods.[40] But what is natural is a relative term, socially determined and often highly contested.[41] From the early decades of the twentieth century, anthropologists and sociologists such as Claude Lévi-Strauss, Mary Douglas, and Pierre Bourdieu have argued that taste and diet are shaped by culture and controlled by society.[42] More recently, historians too have argued that taste, eating habits, and other lifestyle expressions are shaped by economic, social, and political forces.[43] Madeleine Ferrières has shown that fears surrounding food adulteration are not only a product of industrialization, but that food fears, fads, and taboos are constantly present, constantly changing, and largely culturally constructed. Ferrières undermines historical myths that modern society has formed about food, refuting, for example, the modern misconception that food in the past was "more natural" than today.[44]

However, in the nineteenth century, shifting lifestyles associated with urbanization and industrialization led to rapid changes in the ways food was produced and consumed, causing considerable anxiety among contemporaries.[45] Indeed, commentators and novelists in this period often used descriptions of changing food consumption as metaphors for anxieties surrounding an unstable society.[46] These are themes also explored by historians examining the rise of food fads and the pure foods movement at the turn of the century, amid general concerns about industrialization and the degeneration of society.[47] But while some commentators viewed the industrialization of food with horror, others saw the strengthening relationship between science and food as a democratizing and beneficial development, bringing cheaper, healthier food to an increasing population.[48] These questions and doubts as to what constituted food improvements or food adulterants, including the use of additives to color food, were exploited by consultant chemists who maneuvered themselves into becoming key players in new arrangements of assessing and controlling food quality in the Western world. However, the lack of certainty surrounding the dyes, coupled with growing public concern about their safety, proved a difficult dilemma for public analysts, who were fighting hard and successfully to take on the public role of policing food adulteration while at the same time promoting chemistry as society's most useful science. Phillips and French argue that analysts were able to use regulation and food adulteration to create an institutional base in the form of the Society for Public Analysts and to further their careers, reputation, and profile, being employed by both the food industry and public authorities. Food adulteration issues raised the profile of chemistry, enhancing the status of public analysts within private, public, and government spheres, as well as encouraging the formation of consumer and producer pressure groups.[49]

Local municipalities in Britain were among the first state authorities to appoint chemical analysts to monitor the nation's food supply. Indeed, concerns over the use of toxic mineral dyes in food prompted politicians in Britain to introduce food legislation in the 1860s. However, British public analysts were reluctant to take action against the new coal tar dyes being used as food additives. An understanding of the social, political, and commercial context of public analysts' work and status helps to explain the stance taken by these chemists.

The creation of experts, expertise, and expert systems

Concerns surrounding food quality and integrity led to the creation of new expert knowledge systems across Europe and the US, based on changing chemical understandings of food. Examining the public analysts' response to the unmonitored introduction of new indeterminable substances into the food supply is also particularly pertinent to the emerging scholarly research into risk and environmental health. This growing body of work focuses on the interaction between scientific experts, the public, producers, retailers, journalists, and politicians concerned about the management and regulation of industrial and scientific processes and products that may have an adverse impact on health and the environment.[50]

According to Anthony Giddens, the creation of expert systems marked an acknowledgment that many risks in life are a result of humans, not gods or Fate, and a way of providing trust and faith in a product or service where the consumer is removed spatially and in time from the providers or designers of the product or service. Giddens and other commentators on risk, such as Ulrich Beck, argue that the change in risk perception from those risks arising in nature to those created by humans' altering and tampering with nature was a late-twentieth-century phenomenon.[51] However, awareness of the risks of human tampering with food existed long before the twentieth century; the nineteenth-century legitimization of the industrial food supply, including the introduction of man-made substances such as chemical additives, is an example of the creation of an "expert system."[52] Science and knowledge pertaining to the food system were as subject to change and differing expert opinions in the nineteenth century as they are today. Expert systems for managing and controlling changing environments surrounding the production and consumption of food were formed as a result of compromises between different communities.

Chemists acted as a crucial link between food producers, consumers, and governments. The historian Uwe Spiekermann argues that changing chemical and nutritional knowledge helped redefine food in the West between 1880 and 1914, with new expert systems, public trust, and standardization based on chemical definitions of food.[53] Pierre-Antoine

Dessaux, however, points out that chemists found it hard to win trust, claiming that producers and retailers remained the deciding voices in food market regulation. Ximo Guillem-Llobat also has shown that commercial disputes and private sector lobbying played a more decisive role than science and scientists in affecting the outcome of food legislation in Spain in the late nineteenth and early twentieth centuries.[54] This book will seek to assess the role played by chemists in our changing relationship and understanding of food by comparing chemists' response to the use of chemical dyes in food in different countries.

Even when consensus can be reached among scientists in the laboratory, the application of new science in the commercial and public sphere dramatically increases the cultural, economic, and political variances and the number of vested interests, and requires a new approach to consensus building. Analysts were often employed by both prosecution and defense in food adulteration court cases as well as being recruited by manufacturers, lobbyists, politicians, and consumer groups to further their aims and interests.[55] Different interest groups manipulated science to further their own agendas, resulting in a public and media perception of scientists as both poachers and gamekeepers. Synthetic coloring, found in a natural waste product, created in the laboratory, manufactured in a factory, and incorporated into food, society's most basic commodity, is an excellent medium through which to explore the mediation necessary to accommodate, relate to, and understand new substances. As part of the arbitration process, politicians, analysts, and manufacturers all invoked the consumer. While food producers argued that their consumers were well equipped to make their own judgments on food quality and price, food reformers claimed consumers needed protection from a profit-oriented marketplace.[56] Artificially colored food products were becoming increasingly manufactured commodities, widely advertised and available to a wider range of people. Debates surrounding the benefits and risks of novel mass-produced substances, capable of transforming the choice and availability of both basic and luxury food items for all sectors of the population, show how the demand for, and acceptance of, these new products was determined by suppliers as much as consumers. The commodification of new chemically colored food also formed part of the transformation of science and scientists within the public sphere during the late nineteenth century. During this period, chemists were establishing themselves as academics, industrialists, and consultants. On the one hand, they were discovering and manufacturing, on an industrial scale, new products that would transform society, while on the other hand they were being paid as public health advisers

to police issues afflicting industrialized societies, such as food adulteration, environmental pollution, sanitation, and the water supply.[57]

Chemists, consumers, producers, retailers, and politicians all played
a role in the creation of an expert system of trust surrounding industrialized food at a time when new ingredients were becoming legitimized
in food production, a system that is still being challenged today. In the
twenty-first century, an age when the products and processes of science
and technology are being increasingly questioned, a broad and in-depth
understanding of how scientific, technological, commercial, political,
and cultural knowledge are formed and interact with one another is
crucial. This is achieved by following science out of confined cultural
and physical spaces into much wider spheres of operation.

While investigating such an extensive geographical and disciplinary range produces its own challenges and imperfections, it reinforces
claims that the disciplinary and methodological boundaries defining the
history and sociology of science for much of the twentieth century need
to be discarded to better understand social order, knowledge making
and expertise, and credibility in science.[58] By following a scientific creation out of its laboratory into the stomach of consumers in four different nations, this book provides further evidence that many twentieth-
century methodologies of the history of science, which differentiate
between science and technology, pure and applied science, and internal
and external histories, are inadequate to deal with the adoption and
adaptation of scientific and technological objects in society.[59] Tracing
substances outside of the laboratory and into circulation as commercial
objects enables us to say more about the spaces from where they came
and their producers.[60]

1

Food adulteration and the rise of the food chemist

If you opened a can of baked beans to discover a brown gloopy sauce containing brown haricot beans, rather than the orange-red sauce and beans you were expecting, what thoughts would run through your head? Is the can punctured or past its sell-by date or is it a different brand—inferior and cheaper, or possibly organic and more expensive? Maybe the beans were processed in a country with different food regulations or practices? Why should brown beans look less appetizing than red ones anyway? Such thoughts suggest inherent consumer awareness of why food manufacturers color food: to make their products look fresher and more appetizing, to create brand recognition, to maximize profits, to meet consumer expectations, or to comply with national or international regulations.

In many ways, dyes are as important as flavorings in determining how we taste food, with experiments showing that food dyed with different colors, such as apple puree dyed red, changes flavor perceptions. As a result, loss of color in food processing and the subsequent use of artificial coloring to make our food look fresher, more consistent, or even novel is a major issue for food manufacturers.

Indeed, the color of our food plays a central role in our cultural and psychological understanding of food and experience of consumption, with color preferences varying with our age and upbringing.[1] When Burger King launched a black burger in a black bread bun in Japan in 2012, and even blacker versions in 2014, using bread colored with bamboo charcoal and ketchup blackened with squid ink, American commentators cried "gross" and "crazy." Black was not a color associated with food in the US, unless it was burnt![2] Yet black was an acknowledged food color within Japanese culture and so was accepted more easily by consumers there. But within months, the alleged health benefits of activated charcoal led to black bread becoming the latest health fad globally, including in the US.[3] Consumers expect their food to be a certain color, whether that color indicates freshness, novelty, naturalness, or familiarity. However, the presentation of new foods alongside medical or scientific claims can alter long-standing culturally formed ideas about what food should be.

Consumers in the twenty-first century seem to be constantly bombarded with claims and counterclaims surrounding the health benefits or otherwise of food ingredients and additives, much of it related to the coloring of food. Eat red, eat green, eat purple! While science is invoked to justify the claims of food manufacturers and health advisers, consumer advocates and organic food movements scorn the industrialization of food. A return to more "natural food" free of "chemicals" is demanded. Yet all food is made of chemicals. So who determines which chemicals are "natural" and which are "edible"? Visit any food trade exhibition and you will experience the contradiction of dozens of food companies promoting natural food products and ingredients—"free from chemical additives"—alongside dozens of manufacturers exhibiting new "functional" food products containing a multitude of chemicals and other additives included to promote health.

Consumers' confusion is demonstrated clearly in the oft-heard phrase "full of E numbers," a pejorative term used in Europe to describe supposedly "unhealthy" manufactured food full of harmful chemical additives.[4] However, E numbers, which are ascribed to substances extracted from turmeric, saffron, salt, citric acid, and yeast as well as those originally extracted from coal tar and now from petroleum, such as aniline and azo dyes, are only assigned to food additives deemed to be harmless to eat.[5] It is ironic that mid twentieth-century attempts to increase the safety of, and trust in, food, through legislation ensuring that all food ingredients must be approved, scientifically named, and included on food labels, have made consumers more concerned, not reassured, about the safety of their food.

This book describes how colors synthesized from coal tar began being used in food from the mid-1850s and how they became legitimized as food ingredients, although manufactured primarily as textile dyes. While concern about their use was raised in the nineteenth century and continued through the twentieth century, many of the coal tar dyes continue to be recognized as food ingredients despite ongoing legislation restricting their use. Research linking six specified coal tar dyes with hyperactivity and allergies in children in 2007 led to EU regulations making labeling compulsory for these particular dyes. Meanwhile, chemists' and manufacturers' attempts to replace the coal tar colors with "natural" colorings extracted from plants are also now attracting regulators' and consumers' attention, as complex processes involving the use of chemical solvents are required in their manufacture.[6]

Chemists' claims of authority on food and food ingredients are not a new phenomenon. As early as the sixteenth and seventeenth centuries, chemists and doctors were representing themselves as experts in food safety and nutrition. Sixteenth-, seventeenth-, and eighteenth-century chemists including Robert Boyle, Carl Wilhelm Scheele, Antoine Fourcroy, and Herman Boerhaave all wrote about the issue of food adulteration. Meanwhile, food analysis was one of the most influential early aspects of organic analysis, conducted by chemists across Europe including Nicolas Lemery, Heinrich Diesbach, Andreas Marggraf, Caspar Neumann, Antoine Lavoisier, Joseph Louis Gay-Lussac and Louis-Jacques Thénard, Jöns Jacob Berzelius, Thomas Thomson, and Michel-Eugène Chevreul, among others.[7]

From the seventeenth century onwards a markedly different alimentary regime developed in Europe, with a noticeable rise in the number and variety of experts and knowledge claims surrounding food, particularly related to scientific and medicinal understandings of nutrition, chemistry, and physiology.[8] As the consumer became increasingly distant from both food producers and food experts, the question of authority over eating became more significant and contested. Government oversight of national diets increased as wider sections of the population were exposed to new food products, ingredients, and processes resulting from increased mercantile and scientific exploration and factory production.

Chemists from the seventeenth century onwards became increasingly involved in food innovation, often with government support, as well as playing a growing role in assessing and analyzing food.[9] Chemists devised new ways of preparing and preserving food and making unappetizing products more palatable. New foods, from potatoes to cocoa and sugar, were manipulated and processed to appeal to the European palate. Food production no longer remained the preserve of the farmer or artisan. Indeed, by the nineteenth century, chemists in France had already experienced considerable success in staking their claim as experts over the understanding of nourishment.[10]

While chemists constantly sought new commodities, ingredients, and processes in order to nourish the public, they also sought to protect the public from the dangers lurking in the food supply due to fraud and ignorance. It was no coincidence that antiadulteration pamphlets increasingly distributed from the late eighteenth century also raised the profile of chemists.[11] Chemists such as the German-born, London-based Friedrich Accum saw adulteration as a means of raising their profile and status as well as that of chemistry. Adulteration and its investigation thus became part of a wider campaign by analytical chemists to find a more central position in a chemical world where academic chemists, doctors, and pharmacists contended with lower-status practicing chemists and druggists.[12]

Analytical chemistry and the increasing ability of practitioners to detect adulteration in food and water began to take center stage in battles between pharmacists on the one hand and chemists and druggists on the other. Pharmacists tended to be licensed and qualified as physicians and/ or chemists able to diagnose and recommend remedies. Chemists and druggists, meanwhile, manufactured and retailed drugs, often in a shop. These entrepreneurial practitioners were often accused by pharmacists of undercutting drug prices. During the early decades of the nineteenth century, pharmacists in Germany, however, took advantage of developments in analytical chemistry and increasing public concern over adulteration of food, water, and drugs to boost their status and commercial position, during a period when the large-scale production of generic drugs, sold by chemists and druggists, was undermining the pharmacists' specialist role in the marketplace. Changes in chemical techniques towards an increased use of reagents and chemical reactions led to simpler, less expensive, and portable laboratory equipment, allowing more people access to analytical chemistry. This development also created a platform for large-scale chemical training, by making the use of laboratory equipment and practical experiments more widespread and affordable within both mechanics' institutes and universities and other educational institutions throughout Germany, France, and Britain, leading to the emergence of career chemists in the nineteenth century.[13]

Chemical analysis techniques developed by chemists such as Bergman, Gay-Lussac, Berzelius, and Fresenius enabled chemically trained pharmacists to assess the purity of drinking water and detect food adulteration, with concerns about public health and the purity of food and drugs providing pharmacists with an opportunity to reinstate themselves in the marketplace.[14] This was a trend happening across Europe, as evidenced in works such as Jean Baptiste Alphonse Chevallier's *Dictionnaire des Alterations et Falsifications*, published in Paris in 1850.[15]

Science-based expert systems played an increasingly important role in the public's understanding and relationship with food, thanks in part to the growth of the professional and middle classes in the nineteenth century. Men of science were increasingly able to earn a living through the application of chemistry, physics, or engineering in academia, industry, or public service.[16] Chemists across Europe, from Humphry Davy and Michael Faraday to Joseph Louis Gay-Lussac and Justus von Liebig to Louis Pasteur, earned money as well as public reputation through both their academic work and their role as advisers and consultants to industry and the government.[17] By 1842, a guide to professions in France included ten pages describing different types of chemists, while in Germany the author Friedrich Schödler noted in 1875 that "a veritable army of chemists practises their profession with restless energy . . . where there were some scores before, there are hundreds now."[18]

As the public and government increasingly turned to science to determine and manage the perceived new risks of their rapidly industrializing and urbanizing society, scientific "experts" were called upon to assess and improve living conditions, from food production to sanitation to health care.[19] This was a period when the health and welfare of the population was fast becoming a political issue. Doctors and chemists advocating health reform were part of a wider set of middle-class reformers, professional members of a social and political hygienic crusade to improve the sanitation, housing, nutrition, and education of the working classes. Increasing concern about the living conditions of workers and the diseases and lack of hygiene in the industrial cities, publicized by campaigners such as Edwin Chadwick, led to political action in the form of legislation, such as the 1832 Poor Law and the 1848 Public Health Act in Britain, and a call for more experts to advise government in the better regulation of sewage, welfare, hospitals, water supply, food, and the control of disease. The anthropologist Mary Douglas argues that a society's concern over concepts of pollution and dirt reflects its need to create and maintain social order, with pollutants representing the boundary sites between order and disorder.[20] While food adulteration never gained the political visibility of housing, clean water, and sewage, the growing sanitation movement in the mid-nineteenth century provided momentum for government intervention into food regulation and oversight, helped by chemists such as Arthur Hill Hassall, who in 1855 described adulteration as an issue of extreme national importance.[21] At the same time, chemists across Europe were expanding knowledge of the chemical composition of foodstuffs and its effects on the human body.[22]

Mineral and metallic dyes in food cause growing concern

By the nineteenth century, the artificial coloring of food with mineral and metallic dyes became one of the most prominent, certainly the most visible, areas of concern and government action in Europe. As early as 1830 in France, Orders of the Préfet de Police of Paris forbade the

> use of any mineral substance for colouring liqueurs, bonbons, sugar plums, lozenges, or any other kinds of sweetmeats or pastry. No other colouring matter than of a vegetable nature shall be employed for such purpose, except gamboge and archil. It is forbidden to wrap sweetmeats in paper glazed or coloured with mineral substances. It is ordered that all confectioners, grocers, and dealers in liqueurs, bonbons, sweetmeats, lozenges etc. shall have their name, address, and trade printed on the paper in which the above articles will be enveloped. All manufacturers and dealers are personally responsible for the accidents which shall be traced to the liqueurs, bonbons, and other sweetmeats manufactured or sold by them.[23]

The use of substances such as arsenic, red lead, and the mercury-based vermillion to color food and drink had been highlighted in pamphlets such as Friedrich Accum's 1820 *Treatise on Adulterations of Foods and Culinary Poisons*. In 1830, an article in *The Lancet* highlighted the use of minerals, such as the arsenic-based Schweinfurt green, in French confectionery. Subsequent investigations indicated a similar use of toxic minerals to color British sweets.[24] Thomas Wakley, the editor of the *Lancet*, translated Chevallier's antiadulteration work into English, and published *Poisoned Confectionary* by W. B. O'Shaughnessy in 1831.[25] Wakley, a campaigning doctor and Member of Parliament, subsequently commissioned Hassall, a medical chemist and microscopist, to run an enquiry from 1851 to 1854 into the issue of food adulteration. One of Hassall's first analyses was of confectionery, where an investigation of 101 sweets revealed colors created from lead chromate (yellow); red lead and cinnabar (red); indigo, Prussian blue, aluminum, and sodium (blue); Van Dyke brown, umber, sienna (brown); Brunswick green and Prussian blue mixed with lead chromate (green). Such mineral and metallic dyes were used extensively in art and material decorations and the dyeing of fabrics.[26] The problem chemists faced was that colors in food were created by mixing known and unknown pigments, and ingredients of pigments, some of which were poisonous in themselves and some of which could be poisonous in combination with other substances. However, the main concern surrounded the deceptive practice of false coloring.

By the 1850s, *The Lancet* was running a series of articles on the investigations on food adulteration led by Hassall, who would spend his career working on public health and food safety issues.[27] In an 1854 report on the use of minerals to color foods, Hassall identified many toxic materials such as lead chromate, lead oxide, copper arsenite, and Scheele's green and other arsenical compounds. Hassall described food coloring as "very prevalent" and the worst kind of adulteration in the marketplace, as so many of the mineral dyes employed were toxic.[28] Hassall believed that the addition of coloring matters to improve food appearance and conceal other forms of adulteration was "a very prevalent adulteration, and it is the most objectionable and reprehensible of all, because substances are frequently employed, for the purpose of imparting colour, possessing highly deleterious and even in some cases poisonous properties, as various preparations of lead, copper, mercury and arsenic."[29] Among the colors identified by Hassall were salts of copper added to bottled fruits and vegetables; Venetian red added to anchovies and chocolate; chromate of lead, chrome yellow, and turmeric used in custard powder; red lead in curry powder; and many kinds of coloring in confectionery produced from iron, copper, and chromium. While we now know that iron and copper are nutritionally needed, the introduction of such substances to alter the color of food was considered by chemists such as Hassall to be fraudulent and possibly toxic because it was impossible to test the quantities employed and the tolerance of the human body.

Hassall's concerns embraced both loss of income to the state due to lower food costs and deception of the public with subsequent loss of faith in the commercial system. "It is clear," he observed, "that the sellers of adulterated articles of consumption, be they manufacturers or retail dealers, are in a position to enhance their profits by the practice of adulteration, and are able to undersell, and too often to ruin, their more scrupulous and honest competitors."[30] He also pointed to the sanitary and health issues surrounding the use of some of the coloring materials. While admitting that many of the adulterations were safe in single doses, he warned of a cumulative effect of, for example, the consumption of coppers in pickled and preserved fruit and vegetables as well as the occasional fatal single doses particularly in confectionery, noting that "scarcely a year passes by but several cases are recorded of poisoning by articles of coloured confectionery."[31]

Of equal concern to Hassall, however, was the wider moral issue and the importance of trust:

> Adulteration makes not only those who practice it dishonest, but other very serious evils ensue. Thus, it begets the greatest mistrust on the part of the buyer, who loses confidence in those with whom he deals, and thus in this way sometimes the honest trader comes to be looked upon with the same suspicion as the adulterating merchant, manufacturer or tradesman [and] the character of the whole nation suffers in consequence of the prevalence of adulteration.[32]

Hassall's rhetoric was not confined to the pages of *The Lancet* and other journals aimed at the new professional men of science. Extracts from his report were published across the full breadth of the Victorian media, appearing in mass circulation newspapers both in Britain and overseas. *The Lancet's* articles and investigations were widely reported throughout the popular press, both in the regional and London newspapers and in journals aimed at female and male readers. Throughout the 1850s, newspapers and journals recounted many instances in which children and adults were poisoned, often by toxic mineral colorings used to decorate cakes.[33]

The adulteration of food with additives used to make products look more appetizing, or to disguise the fact they had been bulked out or diluted with cheaper ingredients, became a huge issue of public and political concern in the nineteenth century. The mass movement of workers from the countryside into large industrial settlements resulted in the public becoming less aware of the provenance of their food supply. Increasing competition and food handlers in the supply chain incentivized suppliers to use cheaper ingredients to improve their decreasing profit margins. Confectionery became a particular focus in media stories about food adulteration for several reasons, including the fact that the victims of the adulteration of sweets tended to be children. In addition, sweets were often sold by chemists and pharmacists, increasing the likelihood of potentially toxic ingredients being added, either deliberately or accidently.

In 1858 *Punch* published a cartoon showing Death as the "Great Lozenge-Maker," making sweets beside a barrel of arsenic (figure 1.1).

Adulteration and the deception of the public were themes also tackled by novelists, including Charles Dickens, whose Mr. Spenlow in *David Copperfield* pays "honour to the soil that grew the grape, to the grape that made the wine, to the sun that ripened it and to the merchant that

FIGURE 1.1 "The Great Lozenge-Maker." *Punch*, November 20, 1858. Author's collection.

adulterated it."[34] Charles Kingsley, in *The Water Babies*, specifically attacks the poisonous colorings used in sweets. Melanie Keene has shown how Victorian novelists writing for both children and adults frequently referred to developments in science and social issues.[35] Kingsley's novel, like those of Dickens and other Victorians, was originally published in installments in the periodical press. Written between 1862 and 1863 for the monthly literary periodical *Macmillan's Magazine*, *The Water Babies* described how:

> foolish and wicked people make trash full of lime and poison-
> ous paints, and actually go and steal receipt out of old Madame
> Science's big book to invent poisons for little children, and sell
> them at wakes and fairs and tuck-shops. Very well. Let them
> go on. Dr. Letheby and Dr. Hassall cannot catch them, though
> they are setting traps for them all day long. But the Fairy with
> the birch-rod will catch them all in time, and make them begin
> at one corner of their tuck-shops, and eat their way out at the
> other; by which time they will have got such stomach-aches as
> will cure them of poisoning little children.[36]

Similarly, the literary scholar Rebecca Stern notes that Christina Rossetti's allegorical poem "Goblin Market," written in 1859 and published in 1862, also made reference to the prevailing concern over food adulteration and was part of a large Victorian literature commenting on poisoning, market corruption, and food adulteration.[37] In Rossetti's poem, a young girl enters the goblin's market to be tempted with wonderful-looking food that leaves her ill and wasted. Stern claims that adulteration of food, a commodity that is literally consumed, is the most graphic and powerful illustration of commercial corruption in the marketplace.[38]

Food adulteration, with particular concern around toxic metallic colorings, became a dialogue between social commentators, the public, politicians, food manufacturers and retailers, and chemists. In 1855 a British parliamentary inquiry into food adulteration was instigated, followed by an Act for the Adulteration of Articles of Food and Drink in 1860, later superseded by the 1872 Adulteration Act, which led to the appointment of public analysts, municipal chemists who were tasked with exposing and preventing the adulteration of food for profit or harm.

Just as Accum and his fellow pamphleteers tried to set out their stall as arbiters of a reliable trade in food, so Hassall and a new breed of consultant chemists were able to use the public fear of adulteration to increase their status and trustworthiness in an age when the jobbing chemist still struggled to make a living. Hassall helped spearhead a professional class of chemists and doctors focused on improving the safety of food and drink. Significantly, at least two future presidents of the Society of Public Analysts, Alfred Henry Allen and Otto Hehner, trained as assistants to Hassall.[39]

By the time the new chemical dyes entered the food supply in the 1860s, food adulteration, particularly in the form of artificial coloring, was a major social and political concern. Chemists across Europe had identified the "sophistication" of food as a means of enhancing their status and were positioning themselves as experts able to analyze food and drink and act as arbiters of an increasingly complex food supply system. As millions of workers across nineteenth-century Europe flocked out of the countryside into cities, increasingly complex, competitive, and lengthy food supply chains led to concerns about food provenance and quality.[40] Food adulteration became a major social issue and consultant chemists maneuvered themselves to become key players in new arrangements for assessing and controlling food quality, leading to the creation of new expert knowledge systems across Europe and the US, based on a changing chemical understanding of food.

The nineteenth century was a period of considerable change in food production and distribution, as well as chemical, nutritional, and physiological understanding. Exotic foodstuffs from the far reaches of colonial empires became more widely available in Europe while, at the same time, industrially prepared food items, processed using chemical techniques or new technologies such as canning, pasteurization, freezing, and chemical additives, increasingly appeared in stores. While this meant more choice and convenience for the consumer, it also led to greater fears and anxieties about the provenance of food and its increasing commodification. As workers migrated to the cities, the food supply chain lengthened, and many consumers no longer knew the producers of their food. Increased competition forced food suppliers to reduce costs by substituting cheaper ingredients. At the same time, as novel food products such as margarine and new chemically manufactured ingredients entered the market, what was regarded as adulteration became increasingly open to interpretation. The new chemical food additives and industrial food processes made the visual inspection and organoleptic assessment of food, based on smell and taste, more difficult. At the same time, nutritional and physiological science was changing understandings of the different food components and their role in health. All these issues provided opportunities for chemists to take a more important role in the mediation of food.

Whether food is adulterated or not is itself a slippery and contested concept, as the difference between a legitimate ingredient and a deceptive one can be debatable and depends upon a complex framework of views and legislation, which in turn depends upon many factors, including disputes between rival groups of knowledge experts and merchants. For example, the introduction of baking powder into bread and cake products in the nineteenth century resulted in a widespread and long-lasting debate in the media between food reformers, men of science, manufacturers, and public commentators, including novelists, as to which additives were a form of adulteration and which constituted an improvement in food quality and hygiene.[41] For much of the nineteenth century, many chemists and health reformers advocated the complete removal of yeast from bread products, arguing that yeast was a destructive and unsanitary adulteration of food, and that bread could be more hygienically and efficiently raised through the use of chemicals.[42] Meanwhile, food producers today take out the fat naturally occurring in yogurt products, replace it with synthetized sugar substitutes and emulsifiers, and market their product as "low-fat" and "healthy" and containing no artificial colorings![43] Substances in foods change their meanings and move in and out of legitimacy, leading to changes in food production. Relationships between science and social constructions, nature, and politics are not straightforward and are often the result of historically contingent, nonlinear assemblages of disparate parts.[44]

Food fears, adulteration, and industrialization

In the last few decades of the nineteenth century, food anxieties centered on industrialization and became associated with fears about the degeneration of society and the individual, leading to calls for healthier and unadulterated food. As assorted evolutionary theories developed across Europe and America, scientists such as Francis Galton and Max Nordau and social commentators and writers from Emile Zola to Thomas Hardy fueled debates on degeneration.[45] In the burgeoning insanitary cities, poverty, crime, alcoholism, immorality, and insanity seemed to be increasing, encouraging the growth of social movements advocating temperance, eugenics and race hygiene, healthy eating, and sanitation.[46] Food reformers believed the degeneration of the working class was directly connected to the increased availability of industrially produced cheap food and drink. New chemical additives and technical food processing created uncertainty as to the freshness and wholesomeness of food. Reformers also argued that the increased affordability of luxury food items such as sugar and white bread diminished, rather than improved, the diet of the already malnourished poor. Hygienists claimed that industrialization of food had resulted in cheaper but nutritionally inferior food, such as processed white flour, where the nutritious protein germ layers had been removed to help preservation. Sugar also had become significantly cheaper as a result of beet sugar refining methods developed during the Napoleonic Wars. Cheap sugar, together with the introduction of baking powders, changed eating habits to such a degree that by 1879, one-third of recipes in a common American cookbook were for puddings and cakes. New distillation methods and the improvement of bottle corks also caused spirit, beer, and wine prices to fall, storage ability to increase and, in some cases, alcoholic content to double. Coffee, or chicory substitutes, and tea also became more affordable and available. Addictive industrialized foods, described by Sidney Mintz as drugs, were becoming the staple diet of modern society.[47]

A rapidly changing society in which individuals were becoming increasingly isolated from their food sources and food was no longer prepared and cooked at home or within the community became a moral issue. Commercially manufactured food became a metaphor for fear of the capitalist and industrializing marketplace, with food products manufactured by invisible men armed with invisible poisons. Sanitary reformers such as John Harvey Kellogg set up health clinics, promoting "biologic living" and new "healthy" food products to improve both the individual and society.[48] Food fads and healthy eating movements were embraced by those able to afford them, and ranged from teetotalism and fasting to vegetarianism and even the chew-chew fad, a campaign backed by celebrities and politicians which encouraged eaters to chew each mouthful of food dozens of times.[49] Campaigners claimed that by controlling one's food intake, one could take back control of one's body and, by extension, one's society.[50]

However, while reformist campaigners rejected the industrialization of food, new science was constantly invoked to back up their arguments. This included the sciences of economics and efficiency; the emergent human sciences of evolution, anthropology, and eugenics; physiology and the new chemistry of food, calories, and metabolism.[51] At the same time, entrepreneurs and chemists such as Alfred Bird used chemistry to make food more "hygienic," establishing products such as baking powder to replace yeast, while industrial food manufacturers employed chemical ingredients, such as aniline dyes and boric acid, to help decrease costs and preserve and standardize food. At the same time as food producers turned to chemists for advice and help, governments in Europe and the US also began to consult scientists and appoint chemical analysts to ensure a safe and honest food supply. The industrial food revolution and changing chemical knowledge not only changed our food, they changed our political, economic, and social relationship with food.

The French sociologist Claude Fischler argues that food is integral to cultural and individual identity.[52] Both he and Ulrich Beck claim that twentieth-century and current day food scares are representative of the modern age of anxiety.[53] However, historians have shown that food fears, fads, and taboos date back centuries, and are culturally constructed, ever-present, and ever-changing.[54] The adulteration debates of the late nineteenth century had their origin in urbanization, industrialization, and increasingly competitive and international markets. Adulteration became a battleground between producers, consumers, the state, chemists, and hygienists. Rules about food adulteration were never completely under the control of any one interest group, with final outcomes dependent on strategic alliances and market structures.[55]

Chemical dyes began to be manufactured on an industrial scale from the mid-1850s just as the international food trade and the demands of feeding large and dense population centers were both growing and at a time when governments and companies increasingly consulted chemists about food production and safety. Thus, the introduction and subsequent legitimization of chemically produced textile dyes as food additives marked a profound change in the West's relationship with food.

But what exactly were these new chemical substances that would revolutionize our food supply?

2

The wonder of coal tar dyes

Oil and ointment, and wax, and wine,
And the lovely colours called aniline;
You can make anything, from a salve to a star,
If you only know how to, from black Coal-tar.
Tar of the Gas-works, etc
"Beautiful Tar: Song of an Enthusiastic Scientist "[1]

According to *Punch*'s ode to coal tar, chemists were able
to create almost anything from coal tar, a waste product of
the gas industry. Among the multitude of synthetic compounds chemists produced were dyes, perfumes, drugs,
sweeteners, flavors, preservatives, antiseptics, and soaps.
After thousands of years in which alchemists had searched
in vain for the Philosopher's Stone, nineteenth-century
chemists appeared to have discovered the elixir of life and
the ability to replicate any natural product through the
transmutation of black dirty tar.[2]

Chemists across Europe focused their attention on to
coal tar dyes after William Perkin inadvertently stumbled
across a new purple compound in 1856 while working as
a young assistant to the German chemist August Wilhelm
von Hofmann at London's Royal College of Chemistry
(RCC). Hofmann's research into coal tar had already led
to the extraction of benzene and toluene and their con-
version into nitro compounds and amines, and he and his

students were becoming increasingly aware of the potential discoveries embedded in coal tar. Perkin, for example, was searching for a synthetic version of quinine at the time of his mauveine discovery.[3] Although Perkin's attempts at finding an alternative treatment for malaria proved to be in vain, he did discover that the leftover sludge from his experiment was capable of dyeing material purple. The discovery in itself was not so surprising; aniline, one of the compounds that Perkin was distilling from coal tar, was first isolated in 1826 by German chemist Otto Unverdorben, through the distillation of natural indigo, a plant long used to produce purple dyes. In 1840 another German chemist, Carl Julius Fritzsche, also obtained aniline (a Sanskrit-based name that he is credited with having coined) by distilling indigo with caustic potash.[4] Meanwhile, the ability of coal tar compounds to produce different colors was already well known among chemists. Indeed, in 1834 the Hamburg-born chemist Friedlieb Ferdinand Runge produced a substance he called kyanol from coal tar that produced a blue color when treated with chloride of lime.[5]

Early histories of organic chemistry portrayed the rapid development of the industrial synthetic dye industry as emerging from the scientific research of academic chemists, such as Hofmann and Perkin.[6] Homburg and Nieto-Galan claim this simplistic explanation partly stemmed from the speeches and writings of nineteenth-century academic chemists seeking to promote their status and the value of chemistry and chemical education.[7] They and fellow historian Anthony Travis have argued that the new colors were initially developed using long-standing and well-established chemical practices, within chemical communities already producing and using dyestuffs extracted from vegetables, animals, and minerals.[8] For centuries before the discovery of coal tar colors, chemists and their alchemist forefathers had "manufactured" color, with dyemaking a central constituent of chemistry.[9] Throughout the eighteenth century, many chemists, from Sweden's Carl Wilhelm Scheele to France's Louis-Jacques Thénard and Michel-Eugène Chevreul, developed cheaper or more effective dyestuffs. These chemists worked in, or had close ties with, dye- and print-making workshops throughout Europe.[10]

The transition from natural to artificial dyes was never an abrupt nineteenth-century phenomenon. The demand for mass-produced textile dyes developed with vegetable dyes such as indigo before the new chemical dyes were manufactured.[11] Meanwhile, elaborate chemical extraction and distillation techniques had always been employed to transform plants and minerals into dyes. In the eighteenth century, chemists and colorists chemically altered "natural" dyestuffs to produce new "artificial" dyes, such as picric acid and murexide. Picric acid, discovered in 1771 by Peter Woulfe, was synthesized from phenol and used as a yellow dye, while murexide was prepared from uric acid extracted from guano.[12] By the middle of the nineteenth century, the colorists working in European dye houses, printing factories, and the textile trade had extensive knowledge of the various analytic and synthetic procedures needed to extract and produce colors from all sorts of materials. These artisanal chemists played an important part in the early growth of the synthetic dye industry.[13]

Perkin's great success was to recognize the commercial potential of this new vibrantly colored substance and set out to manufacture it and market it to the textile industry. Ignoring Hofmann's advice not to leave his college research work, Perkin applied for a patent, left the RCC, raised money from his family to build a factory, and began to manufacture the new purple dye on a commercial basis.[14] Within months of Perkin's accidental discovery of the new purple dye, the popular press was hailing the potential hidden in tar and the brilliance of the chemists in releasing it.[15] In the aftermath of the 1851 Great Exhibition, the Victorian press was eager to promote British science and technology, so the discovery of a new purple dye by a young British chemist proved to be perfect copy for the burgeoning printed media.

Fortunately for Perkin, the color purple had become increasingly fashionable since the mid-1850s. This was prompted by prominent dye houses and textile mills in the French cities of Mulhouse, Paris, and Lyon producing a range of mauve textiles, using dyes made from guano from Peru exported to Europe as fertilizer. Purple textiles had always been regarded as the height of luxury and a color fit for emperors since the Roman times, when Tyrian purple was extracted from mollusks, a time-consuming and expensive procedure. When the Spanish-born French queen Empress Eugénie began wearing the new French mauves in 1857, followed by Queen Victoria at her daughter's wedding in 1858, the demand for purple mushroomed.[16]

Although it took Perkin about two years to turn his discovery into a commercial product, it was hailed by the press as evidence of Britain's supremacy in industry and science as soon as it hit the market. The news of Perkin's breakthrough was reported extensively in newspapers and periodicals, and within a few years "Perkin" had become a byword for British scientific and technological achievement. This is evident in an article in praise of Perkins's [sic] purple published in the journal *All the Year Round*. Published by Charles Dickens and subsequently his eldest son Charles Dickens Jr. between 1859 and 1895, the weekly magazine was a successor to Dickens' earlier publication, *Household Works*. Publishing many of Dickens' and other novelists' work in a serialized form, the periodical also included nonfiction, such as descriptions of new developments in science.

The article describes just how pervasive the craze for purple attire was, giving credit to Perkin. At the same time, it pours scorn on the use of the word "mauve," the French word for "mallow plant," an agricultural source of red and purple dyes. Rather than "mauve," the word "purple" is used throughout the article with frequent allusions to its extraction from coal tar, in a reference to Britain's industrial power and ingenuity. In the piece, purple is a political force as well as a fashion accessory that is enriching the country artistically and economically:

> One would think that London was suffering from an election, and that those purple ribbons were synonymous with "Perkins [*sic*] for hever!" and "Perkins [*sic*] and the English constitootion!" The Oxford-street windows are tapestried with running rolls of that luminous extract from coal tar.[17]

The article paints Mr. Perkin as "a man who has fought his way up through the mysteries of chemistry" and who "tracked the purple out in the products of distilled indigo, grasping the secret from amidst the red glare and ponderous smoke of an ordinary unenchanted laboratory." Perkin is portrayed as an enterprising, hard-working chemist whose scientific approach had succeeded in achieving something his alchemist forebears could not attain. Words such as "mysteries," "secrets," "spells"—clear references to alchemy—are contrasted with the unenchanted laboratory of a modern scientist such as Perkin. However, the article questions whether chemists have indeed abandoned their alchemist forebears' pursuit of gold. Perhaps the message is that the "gold" now being produced by chemists is simply in the form of tradeable commodities, a rather pejorative allusion to chemistry's commercial dependency. For "never does [chemistry] work so hard and with such staring, eager eyes and acid-stained fingers as when it works at the bidding of trade; commerce being, as we have at last discovered, the special ambition and object of England as a nation."[18] The often-derogatory Victorian view of trade and commerce is another problematic and ambiguous subject in which chemists found themselves embroiled.

Interestingly, while the article praises the commercial potential of chemistry's art, it takes an ambivalent view of chemists' failure to provide cures for disease instead of other more profitable ventures. While chemistry is "performing all sorts of juggling tricks; it is brewing poisons and searching dead man's stomachs for poisons it has invented; it is watching artificial digestion in artificial pouches; it is doing all over the world, simultaneously, thousands of useful, dangerous and curious things," the article notes, adding, however, that it has never "discovered how to stop the death-flood of cholera, the sloughing throat of diphtheria, or the new plague of London now seething in the Thames." The rhetoric of poison and health, and the dangers and competing claims of chemical expertise and physicians' epistemological knowledge of disease, were important and enduring tropes from earlier generations that survived in the Victorian press. Reference to the chemists' failure to clean up the River Thames alludes to a heated debate in the press during this period. As historians have shown, water quality was a contested issue, with public disagreements between chemists and physicians over how disease was caused, cured, and prevented.[19] Chemists such as Hofmann worked for London's Metropolitan Board of Works and other local authorities, performing chemical analyses of water and offering sanitary solutions to the growing problems associated with urbanization and industrial pollution, at the same time that they were inventing and synthesizing new chemicals. As a result, chemists were portrayed in the press as at once creators, detractors, detectors, and arbiters of the new chemicals that were increasingly becoming part of everyday life.[20] Problems to be negotiated included the slippery question of authority and trust when rival scientific professionals produced contradictory evidence, the lack of certainty in "science," and the impossibility of assessing the long-term health effects of small amounts of apparently toxic substances. This is an early example of the contradictions at play between the promises made by scientists and governments in the early twentieth century of "better living through chemistry" and the now recognized, but unknown and unquantifiable, risks arising from the accumulation and synergistic impact of hundreds of thousands of synthetic chemicals used to make our lives better.[21]

These multifaceted considerations did not escape the Victorian public. Knowledge and skepticism among lay readers were juxtaposed with expressions of marvel and optimism over the latest developments in science and technology. The skepticism with which these same new substances were subsequently to be treated in the press indicates that the reading public was far from being the passive observer of science often assumed by historians.

While news reporting to some degree reflected public opinion, how issues were reported in the media also helped form public perception, making information flow a two-way process. Many groups of people, including manufacturers, retailers, professional scientists, and politicians, used the media to further their own interests. Science, food adulteration, public health and hygiene, and government legislation were all covered extensively in the press.

Scientists themselves were keen to publicize and popularize the work of science.[22] A combination of technological developments in printing, a relaxation of fiscal and political constraints on publishing, developments in science and technology generally, and the rise of the middle class and professional scientist, among other social changes, led to an explosion of publications featuring popular science.[23] The second half of the nineteenth century also saw an upsurge of professional and trade journals, including those covering chemistry and the chemical manufacturing industries. Manchester-based dye manufacturer Ivan Levenstein published *The Chemical Review* from 1871 to 1891, while chemical analyst William Crookes founded and edited *Chemical News*. Other trade journals launched in the same period include *The Dyer, Calico Printer*, and *Bleacher and Finisher*, founded in 1879; *Journal of the Society of the Chemical Industry* (1882); and *The Chemical Trade Journal* (1887).

Meanwhile, members of new professional societies such as the Society of Chemistry and the Chemical Allied Industries of Great Britain discussed scientific developments at local branch meetings and events throughout the country. Local chambers of commerce and philosophical societies also hosted debates on science and technology, while new products were displayed at local and international exhibitions.[24] Reports from these journals, societies, meetings, and exhibitions were published throughout the popular press. Scientific developments were covered in news articles, editorial comments, light-hearted cartoons, and exchanges of letters. New scientific discoveries were being talked about and demonstrated in exhibitions, publications, shops, and homes and on the street, helping science to find a wider public audience as well as boosting the social status and authority of scientists.[25]

Science was central to Victorian culture and politics, with novels full of characters caught up in "scientific pursuits."[26] Periodicals also played an important role in shaping the Victorian public's understanding of new discoveries in science, technology, and medicine, using a broad spectrum of literary forms from satire, humor, and fictional tales to serious scientific and news articles.[27] As science became more professionalized, editors began to employ journalists, rather than professional scientists, to write on scientific subjects in order to attract and engage a broader spread of readers including children, teenagers, women, and the working class.[28]

The press played as important a role as chemists and manufacturers in the public's understanding of what the new dyes were and what they represented. The proliferation of journals and newspapers competing for a growing literate middle and working class, at a time when science was regarded as an essential part of cultural life, provides us with a unique opportunity to examine both the general dissemination and the far-from-passive appropriation of science by the public. The general press provided space for a range of opinions to be expressed and debated by the public, scientific practitioners, manufacturers, and politicians. The wide range of publications, and the fact that members of an increasingly broad cross-section of the public would often read more than one newspaper, consumer periodical, or trade or professional journal, created a cultural forum for ideas to be disseminated, debated, and transformed. To date, historians have tended to focus on the most controversial scientific issues debated in the Victorian press, subjects like evolution, astronomy, natural theology, and eugenics, as well as historical themes such as gender issues. However, the press provides us with a unique lens through which to view how the science and technology of everyday commodities came to be debated in a public setting.

Initial mentions of the new aniline dyes in the press suggested a degree of wonder and confidence that resourceful chemists had found innovative materials to create novel and exciting products, improve existing products, and replace dangerous products and foreign imports. This optimism, shared by the press and its readers, is evident in the public and press reception to events such as the 1862 International Exhibition, in which the industrial and scientific prowess of Britain was linked to its position at the head of a global empire.

The rhetoric used in the press during the 1860s of coal tar organic chemistry was based directly on the public communication of the scientists themselves, principally Hofmann. Hofmann was an avid popularizer of chemistry who, as director of the Royal College of Chemistry (RCC), advocated a role for science in education, industry, and the economic and social well-being of a nation. While at the RCC, Hofmann taught many of the individuals whose work would lead to the formation of the modern organic chemical industry, including Edward Nicholson, Carl Martius, and Johann Peter Griess. He himself was responsible for the discovery of other dyes, such as Hofmann's violets.[29] In a lecture given at the South Kensington Museum to science teachers, Hofmann described "benzol," the base material of many of the new dyes:

Capable of undergoing in the hand of the chemist an endless number of Proteus-like transformations, this substance has not only assisted the progress of science itself, but given rise to new and important branches of industry. . . . in perfumery and confectionery we use it as a substitute for the essential oil of bitter almonds.[30]

This speech and others, such as Hofmann's talk on "mauve and magenta" presented to the Royal Institution, were reported widely in newspapers and periodicals, and their tone and content is clearly reflected in subsequent articles on the new dyes in the popular press.[31]

In April 1861, *The Ladies' Treasury*, one of a cluster of new periodicals launched in the mid-years of the nineteenth century published specifically for middle-class women, announced that "benzine, a product of coal-tar, is to become of manifold use," from killing insects and rats to its employment, when combined with nitric acid, as an almond flavoring in confectionery, and in custards and blancmanges. Meanwhile, "when mixed with zinc filings, and weak sulphuric acid, [it] is transformed into aniline, a colourless oily substance, which, when oxidized by chromate of potash, is transformed into mauve or lilac dye."[32]

The extensive use of chemical terminology adopted in the article suggests that its middle-class female readers sought to educate themselves about science and what it could do for them.[33] Orland has described how chemists such as Friedlieb Ferdinand Runge, Justus von Liebig, and Hofmann were prominent among nineteenth-century chemists keen to write about chemistry and its application in everyday life for a wide public readership.[34] Liebig and Hofmann's work, especially, was subsequently published in newspapers and periodicals, and from the 1860s we can clearly see discussions of the practicalities of chemistry spreading throughout a wide range of Victorian publications, including newspapers, women's journals, children's magazines, and comic and satirical periodicals, suggesting a growing public demand for more scientific knowledge. Chemistry, particularly during this period, was producing visible and practicable products, like the new dyes, which made eye-catching displays at the increasingly popular exhibitions and trade fairs.[35] Describing the "exquisite dyes produced from coal" in an article on the 1862 International Exhibition, *The Ladies' Treasury* waxed lyrical about the array of colors released from "that imprisoned life which thousands of years long past was encased in decaying vegetable substance—[that] has at last sprung into light and beauty, at the magic touch of science."[36]

This image of trapped sunbeams and imprisoned life, released after millions of years by scientists and brought back to delight and provide service for an industrial world, was one that was painted across the press. In 1865, another article in *The Ladies' Treasury* shows that, while the sense of wonder had not diminished—it described the "brilliant hues which our modern chemists have eliminated from the 'black stones'" as "the imprisoned sunbeams of ages ago"—there was a recognition of just how widespread the use of coal tar products had become, and of the democratizing aspect of chemistry in bringing color to the masses. The article noted that "the brightest colours that deck the form of beauty, or that lend a charm to the peasant maiden, are obtained from coal."

While stressing both the usefulness and the wonder of coal and chemistry, the journal again indicates the level of knowledge and interest its female readers were expected to possess. "Coal, which is composed of but four elementary bodies, viz. carbon, hydrogen, oxygen and nitrogen, when put into retorts, and subjected to distillation, and become decomposed by heat, produces nearly fifty other different substances, every one of these fifty products being of some use," the journalist reported.[37]

Fashion historians note that the British public immediately embraced the new, bright synthetic colors produced by British, French, and German dyemakers as compared with the duller palette of vegetable dyes.[38] At the same time, the growth of fashion periodicals increasingly exposed middle-class women to the latest fashions.[39] According to Alison Gernsheim, during the "crinoline period" of the 1860s, the synthetic dyes of "purple, mauve and the bright purplish-pink or fuchsia shades named Magenta and Solferino in honour of Napoleon III's victories over the Austrians in 1859, were all the rage."[40] The craze for the new purple dye began in France, where the more innovative and technically competent dyers of Paris and Lyon were more eager to try the new dyes than the more conservative British dyers.[41] However, the British soon adopted the French fashions with an exuberance that prompted the French philosopher and historian Hippolyte Taine to comment on the "outrageously crude" colors worn by women in London, "overdone, loud, excessively numerous colours, each swearing at the others." Observing a "ridiculous" spectacle of "shop-keepers'" wives in Hampton Court wearing violet dresses "of a really ferocious violet," Taine felt compelled to comment to one lady that clothes in Britain were more "showy" than in France, only to be told that "our dresses come from Paris!"[42]

Punch, as usual, was quick to remark on the waxing and waning of the new colors as they fell in and out of fashion, as the cartoon in

FIGURE 2.1 "Scene: Commercial Room." *Punch*. Author's collection.

figure 2.1 illustrates. "You're always in the fashion, I see. Last time I had the pleasure of seeing you, Mauve was the prevailing colour, and your Nose was Mauve. Now Magenta is all the go, and it's changed to Magenta."

Adulation or adulteration?

However, it was not long before the amazement with which the popular media had hailed the discovery of aniline dyes showed signs of corroding, as cases of poisonous aniline-dyed socks, hats, and gloves began to be reported widely in the press. A doctor informed a London magistrate that several of his patients had complained of discomfort caused by socks and clothing colored with the new synthetic dyes; this created much debate in the pages of *The Times*, *The Lancet*, and other publications. Members of the public, doctors, public analysts, chemists, journalists, and those who crossed disciplines, such as the editor of *Chemical News*, consultant William Crookes, publicly debated in print the probable causes of skin complaints and other injuries allegedly caused by the dyes. When a Coventry doctor blamed picric acid or aniline yellow for causing skin complaints after reading the German chemist Max Reinmann's book *On Aniline and Its Derivatives*. Crookes, who had translated the book, claimed in a letter to *The Times* that aniline yellow was harmless. Rather, he suggested that it might have been the alkali used with the dye that had caused the problem or, more probably, alternative dyes "Victoria orange" (dinitroparacresol) or "Manchester yellow" (2,4-dinitro-1-napthol), both discovered by Carl Martius in 1864, or even the dinitroaniline chloroxynapthalic acid and nitrophenylenediamine dyes used to color things brilliant red. He offered to identify the poison if samples were sent to him. Other readers related their own stories, with analysts declaring some dyes harmless and pinpointing others as harmful.[43] Initially, however, chemists seemed reluctant to identify adverse reactions with aniline itself. For some years, contaminants such as arsenic, which were used in the production process, were blamed for any problems, or failing that, dyes with which the analyst had no allegiance or experience, such as new and unknown dyes, often ones discovered by a "foreign" scientist.

Despite the initial reluctance of chemists to discredit the new dyes, these beautiful substances transmuted from black and dirty waste by the marvel of science were beginning to be seen as problematic, complex, dangerous, and unnatural. As was often the case, *Punch* led the charge to undermine the credibility and reputation of science's latest miracle, declaring in 1868 that it now knew

> what killed Hercules. The shirt of Nessus was not imbued with the poisoned blood of the Centaur. Of course Deianeira, before she sent it to her husband, washed it out. No doubt the garment was one which had been dyed brilliant red with chloroxy-nitric acid, dinitroaniline, or some or other of those splendid but deleterious compounds of aniline which in coloured socks are blistering the feet and ankles of the British public.[44]

Founded in 1841, *Punch* had a weekly circulation of sixty thousand by 1860. Throughout the nineteenth century, the satirical magazine covered science, technology, and medicine as extensively as any Victorian periodical, placing science in its social and political context and revealing its benefits, dangers, genius, absurdity, and usefulness. Contributors with scientific or medical backgrounds were regulars among the journal's anonymous commentators on science. While *Punch* has long been mined by historians for material and viewed as an important nineteenth-century source, it is only in recent years that historians have

realized the extent to which *Punch* and other periodicals acted as producers, as well as disseminators, of knowledge.[45]

A similar report from the *British Medical Journal*, about a shoemaker from Strettin poisoned by his felt hat, circulated in many periodicals, including an 1875 edition of London's popular weekly *John Bull*.
The hat, it was claimed, "caused the poor gentleman to suffer a severe
headache, and eruption and swelling on his head that proceeded to sup-
puration. The hat was placed in the hands of an official analyst, who
found that the brown leather lining was coloured with an aniline dye
containing poison."[46] *Punch* produced the following cautionary poem.

> We've heard of socks that poisoned feet,
> Hats 'gainst heads are now combining.
> With poison in the four-and-nine,
> Lined with the dye of aniline—
> Death may haunt any linin![47]

Similar comment from the mass-circulation penny comic paper *Funny
Folks* is typical of many newspaper articles that played on the words
aniline or dyeing: "A shoemaker at Strettin has been poisoned by wearing a felt hat, the lining of which was coloured with an aniline dye,"
it noted, adding that, "fortunately, all the dy'ing was confined to the
hat—the man lives."[48] The Strettin hat-poisoning incident continued to
be a subject of press comment for at least two years, suggesting that
tales of poisonous hats and socks had taken an almost apocryphal turn,
becoming a truism and media trope for the dangers of synthetic sub-
stitutions. The frequent mention of the same story and, indeed, often
identical articles in different journals is also an indication of the need
to fill many pages with cheap copy in a burgeoning Victorian supply of
popular periodicals and daily newspapers. It is a colorful example of
Rubery's argument that news became a commercial commodity to sell
newspapers in the nineteenth century and suggests that the issue surrounding dyes was apparently considered newsworthy, either as a part
of the adulteration debate or because of concerns surrounding trust in
chemistry and science or a conjunction of the two.[49]

Across the press, dyeing and aniline became ammunition during the
1870s for puncturing inflated images of science or fashion. A cartoon
in *Punch* in 1877, titled "True Artistic Refinement," was subtitled "Died
of a colour, in aesthetic pain." It showed a gentlemen noting of a young
lady "she affects aniline Dyes, don't you know! I weally couldn't go
down to Suppah with a Young Lady who wears Mauve Twimmings in

FIGURE 2.2 "True Artistic Refinement." *Punch*, February 17, 1877, 66. Author's collection.

her Skirt, and Magenta Wibbons in Her Hair!'" (A lisp was often used in Victorian literature to represent an effete, or overly affected, aristocrat.)

However, the transition in the popular press from portraying synthetic dyes as substances of wonder to viewing them as problematic and complex concoctions was not straightforward. Many articles blamed the danger of these new synthesized substances on rogue contaminants, and not the scientifically discovered substance of aniline itself. Even *The Ladies' Treasury*, which until then had still been praising science for releasing the "trapped sunbeams of ages past," was by 1875 noting that "many of the colours derived from coal-tar are known to possess poisonous qualities, and all of them are looked upon with suspicion by ultra-careful housewives." However, while the journalist noted that one type of aniline green, prepared with piric acid and arsenic, was poisonous, and that the dyes of rosaline and corallin, also often contain poisons, he or she reassured readers that aniline dyes "are all quite harmless when pure." And the author suggested that dyed goods only be sold with certificates stating their freedom from arsenic.[50]

From 1860 many dyes, including magenta, were manufactured by reacting aniline with arsenic acid.[51] However, while popular with dye manufacturers as an effective oxidant, arsenic gradually became viewed by others as a dangerous component of dye manufacturing. An article in 1875 advocated the use of aniline dyes for coloring Easter eggs while warning of the danger of arsenic used in the preparation of such dyes.[52] Later that year, a journalist writing for *The Englishwoman's Domestic Magazine*, using the same classical reference to Hercules as *Punch* had used the previous decade, warned the magazine's middle-class readers that "an Austrian chemist, Professor Ginti, has sounded a note of alarm to ladies. He says that English and Alsatian manufacturers of cotton goods, in order to save the expense of albumen to fix the new aniline dyes, employ arsenical glycerine and the arsenicate of alumina for that purpose. . . . Recollect the fate of Hercules, who put on a poisoned garment, and suffered such excruciating torments that he burnt himself alive to escape them."[53] Again, the author suggested that it was not the "science" that was to blame, but the money-saving guile of the manufacturers. Note also that the Germans complained of toxic dyes coming from Britain, while, as will be seen later, the British press would complain about poisonous dyed goods being imported from Germany!

The trouble with arsenic

Arsenic, it seems, had become the universal scapegoat, blamed for any adverse effects associated with aniline dyes. The historian James C. Whorton has referred to Britain's "Arsenic Century," describing the ubiquitous and often fatal presence of arsenic in nineteenth-century domestic settings. Frequently identified as an ingredient in household goods, introduced either accidentally or deliberately, arsenic was welcomed into the home in products such as wallpaper and candles without its possible toxic effects being foreseen.[54] According to Whorton, arsenic is an example of many substances that "become established in commerce before their dangers are recognized, ensuring that any attempt to curtail their use will be resisted by manufacturers with vested interests in their continuation and fought or ignored by politicians ideologically opposed to government interference with business and/or beholden to powerful lobbying groups." Whorton suggests this is a familiar story that has been retold throughout the twentieth and twenty-first centuries.[55] Indeed, the new coal tar dyes in food were also an example of new substances welcomed into the home, indeed into consumers' bodies, without their possible toxic effects being foreseen. Ironically, though, it was the presence of arsenic itself in many of the dyes that, for a long time, deflected concerns about the chemicals themselves. Concerns about the use of arsenic in dye production also included environmental issues surrounding water and ground contamination.[56]

FIGURE 2.3 "The Arsenic Waltz: The New Dance of Death (Dedicated to the Green Wreath and Dress-Mongers)." *Punch*, February 8, 1862, 54. Author's collection.

Dyes become stained

Media coverage during the 1870s suggests that the aniline colors of the previous decade were becoming less fashionable. Perhaps the incidents of poisoning were affecting their image or possibly it was simply that they were becoming more commonly used, and thus were no longer a novelty and no longer expensive. Even the use of the new aniline dyes in luxury items of clothing, sold by the most reputable of tradesmen, was not enough to protect their flagging reputation, as *The Englishwoman's Domestic Magazine* observed in August 1878:

> Even when our lady friends patronize the best houses, and expend a guinea upon their silk stockings, they are by no means protected from the danger of poisonous dye. A correspondent to the Lancet relates a case which came upon his immediate notice, in which a lady suffered most severely by wearing magenta-coloured stockings, which, although supposed to be of most superior quality, were pronounced by Southall Brothers and Barclay, the Birmingham analysts, to have been dyed with coralline, an aniline compound, which invariably has an injurious

effect upon the skin, owing to poisonous minerals introduced into the colouring matter.[57]

This article, like others around this time, attributed the poisonous qualities of aniline dyes to the arsenic introduced as part of the manufacturing process, rather than being an intrinsic feature of the new substance itself. The quote also indicates that, by the late 1870s, the public was looking to chemists and physicians to adjudicate the use of these new substances in society.

Meanwhile, as *Fun, Punch's* liberal rival, pointed out in July 1873, science was producing yet more new colors. In fact, from the 1870s dozens of new dyes were appearing annually, making color an increasingly fashionable commodity, a phenomenon parodied by *Fun* with a pun on "blue":

it is stated by a scientific journal that "the new anthracene blue is in some respects superior to the well-known aniline blues, but it is at the present time necessarily much more expensive. . . ." [This is] tantalizing to persons who wish to feel blue after the anthracene fashion. The inadequacy of the aniline blue has long been felt a public privation, but now that the other sort has been actually found superior, let us indulge the hope that in time it will become so cheap as to be undesirable.[58]

By July 1876, Samuel Beeton's *The Englishwoman's Domestic Magazine* was imploring its readers to "avoid the harsh magenta reds and the cold, hard violets of the aniline dyes, so common a few years ago. In the best warehouses, however, these vicious tints are not to be found."[59] Within two decades the wonder that had greeted Perkin's sensational substance was being replaced with a skepticism and anxiety capable of producing adjectives such as "cold," "hard," and "vicious" to describe the new dyes. Beeton's monthly publication, published between 1852 and 1877, had a circulation of about fifty thousand in 1857 and a cover price of 2d. It was one of the first cheap periodicals published in Britain for the growing numbers of women with the time and ability to read at home. Filling a gap between the expensive women's monthlies and the cheap weeklies, Beeton's magazine included poetry, fiction, and useful and practical knowledge and instruction, including politics and science.[60] The proliferation of such periodicals in this era indicates the growing levels of education and self-education among both male and female middle-class readers and their desire to access more knowledge.

It is no coincidence that the media fascination with the ever-growing display of eye-catching colors began to fade as Britain's production of the new dyes also declined. Chemical dyes were first manufactured commercially in workshops and small factories in Europe's long-established dye, printmaking, and textile industries, particularly in France and Britain. However, within a few decades of Perkin's discovery, most of the world's chemical dyes were being manufactured on an industrial scale in a handful of large production facilities in Germany, where most of the research was also being done. Among the industrial conglomerates established during this period to produce dyestuffs were Badische Anilin und Soda Fabrik (BASF), Aktien Gesellschaft für Anilin-Fabrikaten (AGFA), Hoechst, and Bayer. These dye manufacturers expanded into pharmaceutical and other industrial sectors as a result of their research into synthetic substances obtainable from coal tar waste. The new substances they created rapidly infiltrated everyday life, in the form of perfumes, flavorings, dyes, and drugs.

By the end of the nineteenth century, there had been a massive rise in the production of synthetic or "artificial" dyes. Production of synthetic dyes increased 3,800 percent in monetary value between 1862 and 1913, and 4,500 percent by weight between 1871 and 1913, from 3,500 to 162,000 imperial tons.[61]

But by 1870, little more than ten years after the first aniline dyes began to be produced in England and France, Germany was producing 50 percent of the world's dyes. By 1900, that figure had risen to 85 percent, while Britain's share of the world dye market had fallen to about 3 percent.[62] Just five German companies controlled more than 90 percent of German production and 76.5 percent of world production.[63] Indeed, German supremacy in the chemical industry was a matter of considerable concern and debate in the British press. As far as the British press was concerned, the position was made even more intolerable because coal tar dyes had been a British invention. "Germany is overtaking us in iron and steel, and threatening us in textiles; but she is beating us hand over hand in chemicals. Our English chemists have for a long time past lived in Queer Street," moaned the *Review of Reviews* in 1896, adding that "what makes it worse is that the aniline dyes were the discovery of an Englishman, and at first the whole trade was in English hands."[64]

Of course, the press overlooked the fact that Perkin had been an apprentice to Hofmann, a German chemist brought over to Britian with the help of Queen Victoria's German husband to help bring Britain's scientific education up to the standard of Germany's. For most of the nineteenth century, German chemists played an important role in the wider dissemination of knowledge, education, and practice of organic chemistry and the acceptance of chemical products across Europe and the US. Overlapping networks of chemists moved between Germany and overseas, particularly Britain, before production of these new coal-derived substances became centered on Germany.

Chemical dyes become the foundation of Germany's chemical empire

While the Victorian press portrayed the early production of synthetic dyes as a British phenomenon, German chemists were a crucial component of Britain's chemical community from the mid-nineteenth century. Britain offered an established manufacturing base in which Germany's chemists could work, while German organic chemists trained many British analytical and manufacturing chemists. In the 1850s and 1860s, several of Hofmann's former pupils founded or worked for dye-manufacturing companies in Britain.

During this time, Britain's main competitor in the dye industry was France, whose thriving and long-standing textile dyeing and printing communities also were experimenting in the production of aniline dyes, resulting in the discovery and production of fuchsine and other dyes.[65] Britain and France were ideal environments for the fledgling chemical dye industry. France was a leading European center for analytical chemistry and had well-established and thriving printmaking, dyemaking, and textile enterprises. Britain too had a flourishing textile industry and a strong manufacturing base, including in chemical production. Both countries profited from extensive empires and access to raw materials, such as indigo from the American and Indian continents.[66] France also had ready access to home grown madder, the source of many red and yellow dyes, while Britain benefited from domestic coal supplies. While Germany too had long-established dye, printing, and textile industries, the newly formed nation's access to colonial trade in dyestuffs and textiles was limited. Earlier attempts in the eighteenth century to impose protectionist measures to fend off the threat of imported indigo dye to domestic woad cultivation had largely failed.[67] What Germany did possess by the middle of the nineteenth century was a strong educational base in organic chemistry. Justus von Liebig's research school in Giessen had attracted students from across the Western world for several decades, and his method of laboratory-based research was replicated by other German chemists such as Friedrich Wöhler in Göttingen, Hermann Kolbe in Leipzig, and Robert Bunsen in Heidelberg as well as Hofmann in London and, later, in Berlin.[68]

While Perkin left the Royal College of Chemistry (RCC) to pursue a career in industry, his German mentor Hofmann remained at the London-based college. Hofmann's role as an educator and consultant to the dye industry was to prove as pivotal to the development of the synthetic dye industry as was Perkin's career move into manufacturing. Hofmann had been appointed principal of the newly created RCC in 1845 on the recommendation of Liebig. Hofmann was a migrant product of Liebig's extensive network of students, having studied chemistry as a young man in Liebig's Giessen-based laboratory, part of which had been designed by Hofmann's architect father. Liebig's work on organic chemistry and its useful applications to industry and agriculture was much admired in England. Indeed, the RCC, established by former British and German-born students of Liebig and promoted by Prince Albert, the German-born husband of Queen Victoria, was an attempt to establish in Britain the type of chemical research laboratory teaching practiced by Liebig in Germany.[69]

Hofmann was credited with initiating a school of investigation into coal tar, seeking to create greater theoretical and practical understanding of the thousands of molecules present in this abundant by-product of the gas industry. His program of experiments on the fractioning of coal tar, conducted with his early assistant Charles Mansfield, led to the discovery of many substances, such as toluene and benzene, which formed the basis of coal tar chemistry in London.[70] German chemists, many trained by Hofmann in London, began working in Britain in dye development and production. Carl Martius discovered Manchester yellow, an aniline yellow dye manufactured by Simpson, Maule and Nicholson, and Manchester brown, a dye made by Roberts, Dale & Co., where Martius had worked before returning to Germany to set up AGFA. Heinrich Caro also worked for Roberts, Dale & Co. in Manchester before returning to Germany, where he helped to discover synthetic indigo and patented alizarin while working for BASF. The research of another German chemist, Johann Peter Griess, who worked with Hofmann at the RCC and later for a Burton-on-Trent brewery, led to the development of azo dyes. Azo dyes are formed by combining diazo compounds (organic compounds with two linked nitrogen atoms) with aromatic amines or phenols. Another key researcher in the field of azo dyes was the former Hofmann pupil Otto N. Witt, who worked as a senior chemist for the London-based dye manufacturer Williams Bros & Co., founded by Greville Williams. After returning to Germany in

1879, Witt developed theories relating color to the structure of organic compounds as well as discovering chrysoidine and tropaeolin and a series of azo dyes. Meanwhile, Williams Bros., later Williams (Hounslow) Ltd., was one of the few dye companies to go on to specialize in the production of dyestuffs specifically for the food industry, in the 1920s.[71]

As the synthetic dye sector grew, British companies were forced to recruit more chemists directly from Germany to reinforce the chemists trained by Hofmann in London. Germany's universities and technical schools produced far more scientists, including chemists, than the universities and higher educational establishments of Britain, prompting numerous complaints by British-based industrialists and scientists, such as Ivan Levenstein, himself an immigrant from Germany, and Lyon Playfair.[72]

Germany turns to science and industry

Germany was formed as a nation-state in 1871 as a result of Prussian chancellor Otto von Bismarck's unification plan. As a new nation lacking a substantial empire, Germany sought to create its own techno-industrial economy to compete with Britain's and France's established manufacturing skills and resource-rich colonial empires.[73] Historians have linked nation-building and nationalism in this period to the development of *Naturwissenschaft*, or science and science-based technology in Germany, describing how science was integral to the new nation's social, cultural, and political unification.[74] As Hugill and Bachmann have argued, "a polity's geopolitical stance critically impacts its development of technologies."[75] University reforms and politically led economic reforms had laid the groundwork for the German region's industrial growth from the early 1800s, while the continental blockade of the Napoleonic Wars had prompted German and French chemists to research substitutes for products such as coffee, sugar, and natural spices and medicines, resulting in a surge of chemical research, especially around alkaloids.[76] Throughout the first half of the nineteenth century German politicians encouraged industrial development by handing government-purchased foreign technology over to private parties, and produced development funds for new industrial start-ups to help build state-financed industries.[77] Politicians also encouraged new types of educational systems centered on research and often based at universities but linked to industry.[78]

Historians have argued that the German state's investment in non-directed academic research during the nineteenth century helped lead

Germany to a rapid growth in scientific knowledge and industrialization.[79] However, its academic research was always, from the outset, also
closely linked with industry. Alan Rocke has pointed out that many of
Germany's theoretical chemists, including Hofmann, were trained in the
practical and applied research based laboratories of Wöhler, Bunsen,
and Liebig.[80] Germany's "industrial revolution" was thus particularly
characterized by an intertwining of academic and industrial knowledge,
making it different from the first industrial revolutions in Britain and
France, in which training and innovations were largely empirical and
industry based.[81]

Germany's strategy produced a large pool of well-trained chemists.
During the middle of the nineteenth century, these chemists welcomed
the opportunity to work in Britain's thriving chemical industry and gain
practical experience. One of the main figures in the German chemical
diaspora across Europe, Hofmann himself remained in London for
twenty years, conducting research into coal tar and synthetic dyes and
acting as a consultant to the nascent British-based synthetic dye industry. However, in 1865, he left Britain to teach chemistry at the University
of Berlin. According to his former student Martius, Hofmann returned
in order to set up a Liebig-style research school, after the Prussian government offered to fund a laboratory with up to one hundred students,
a much larger operation than the RCC's laboratory. With Hofmann's return to his homeland, and the subsequent return of several of his fellow
countrymen, including Caro, Martius, and Witt, the knowledge center
and production base of synthetic dyes migrated to Germany.[82] Germany
established itself as a pivotal knowledge center for organic chemistry,
with many of Britain's leading analytical chemists such as Otto Hehner,
Augustus Voelcker, James Wanklyn, and August Dupré, completing part
of their education in Germany.

Much of the ensuing success of the German dye industry can be attributed to its geographical and social proximity to this growing knowledge base of organic chemistry in Germany's universities and technical
schools. University-based organic chemists such as Hofmann and Adolf
von Baeyer worked closely with Germany's expanding industrial companies. Examination of the key players and companies in the German
synthetic dye industry reveals the close connections in the sector. Martius continued to work with Hofmann after both men had returned to
Germany from Britain, with Martius employing many of Hofmann's
Berlin graduate students at AGFA.[83] Baeyer had studied with August
Kekulé and Bunsen before succeeding Liebig as professor of chemistry
at the University of Munich. Carl Gräbe, who worked at Meister Lucius

und Brüning (later to become Hoechst) before returning to academia, and Carl Theodore Liebermann, who subsequently succeeded Baeyer as chemistry professor in Berlin, also both studied in Heidelberg under Bunsen. Gräbe and Liebermann, working with BASF's Heinrich Caro, were the first chemists to synthesize alizarin, a dye naturally found in the root of the madder plant, and the three men applied for a joint patent with BASF.[84] In 1871, BASF produced fifteen tons of alizarin, rising to two thousand tons annually by the turn of the century. BASF's main German competitors, Hoechst and Bayer, also produced alizarin, using different methods. In 1880, Baeyer, by now relocated to Berlin, derived a synthetic indigo, a dye that went into commercial production following further extensive research by Hoechst, Bayer, and BASF.[85]

This was a community of chemists working in different companies and universities both collaboratively and in competition, to further their own interests but also those of chemistry and of the new German Empire. Aided by Kekulé's work on the structure of benzene and Griess's and Witt's studies of azo dyes, the close network of university-educated organic chemists in industry and academia helped fuel the rise of Germany's synthetic dye industry.[86] The heuristics for azo dyes involved testing apparently limitless combinations of diazo compounds with aromatic amines or phenols to form azo compounds, most of which are dyestuffs. Through this systematic synthesis, German academic and industrial chemists created thousands of new colors.[87] Caro described the manufacture of azo dyes as an "endless combination game" achieved by the application of "scientific mass-labor" (*wissenschaftliche Massenarbeit*).[88]

Links to Germany's burgeoning universities and polytechnics enabled German-based dye manufacturers to recruit the increasing numbers of chemists required to work in their factories and large laboratories. Academics maintained links with industry to help secure postgraduation work for their rising numbers of students and, by helping to support German industry and the economy, academia was able to obtain additional funding from government. The close two-way links between academia and industry were among many intertwining strands in the co-evolution of Germany's science and technology, economy, and culture.[89]

As Bismarck sought to build a technocratic state, German chemical firms benefited from an increasingly favorable commercial and political environment. Successful lobbying of government officials by leading German chemists such as Hofmann and Bayer's director, Carl Duisberg, led to preferential legislation, tariffs, and taxes. Among these was a new German patent law in 1877 protecting new dye processes. Before 1877, German companies and inventors had little patent protection compared to French and British companies, who benefited from existing patent law protection in their own countries. However, the initial lack of a patent law had reduced the entrance barriers to the sector in Germany, leading to more entrepreneurial chemical ventures and a very competitive manufacturing and research climate.

The German 1877 patent law was written with the new chemical dye industry in mind, unlike existing French and British patent, laws which were based on the mechanical technology of an earlier industrial landscape. The older French laws, for example, protected products rather than processes and, by the 1870s, significant resources were being lost in patent battles between firms in France and Britain. At the same time, domestic competition in France and Britain was stifled, as firms were effectively granted patent monopolies. Patents obtained by Lyon-based dye firm Société la Fuchsine prevented other companies from manufacturing aniline-based red dyes in France, contributing to a decline in French dye companies and an exodus of dye entrepreneurs to Switzerland and Germany.[90] Germany's 1877 law was introduced at a time when German companies had already pushed ahead in terms of innovation and efficiency, and gave German inventors fifteen-year monopolies on novel dye processes, rather than individual products, helping to consolidate the strength of the dominant German firms.[91] Moreover, German law required that patents be worked in Germany, an arrangement not applied in France or Britain. By the end of the nineteenth century, dye patent applications in Britain, France, the US, and Germany were dominated by German firms.[92]

Power and politics in German chemistry

Organic chemistry played an increasingly prominent role in Germany's economy; by the end of the nineteenth century, German organic chemists had established themselves as a powerful group of professionals within the unified state of Germany.[93] Within a few years of unification, an array of trade and professional associations had been formed to represent the interests of chemists and chemistry in Germany.

One of the first was the German Chemical Society (Deutsche Chemische Gesellschaft), founded in 1867 by Hofmann to allow theoretical and industrial chemists to exchange ideas and information. Chaired by academic chemist Adolf von Baeyer with statutes drawn up by industrial chemist Carl Martius, with Emil Fischer succeeding Hofmann as director, the DCG became dominated by academic chemists operating in the chemical dye industry.

In the following decades (see table 2.1), associations such as the Verein Deutscher Chemiker (German Society for Applied Chemistry) were established to represent all working chemists, including analytical chemists in the public sector and food chemists, who suffered from low status and pay compared with chemists working in academia and industry. Nevertheless, by the end of the century, leading industrialists and academic chemists such as Hofmann, Fischer, and Duisberg dominated the most influential German trade and professional associations and also sat on several governmental committees. As a result, the status and working conditions of industrial chemists, and to some extent of their analyst peers in the public and private sector, were partly contingent on a small group of leading industrialists and academic chemists who held the reins of the most influential associations and journals and the ears of the politicians.[94] These men helped to set the political and economic agenda in favor of academic chemistry, as well as the chemical industry and its employers, particularly in the field of organic chemistry and especially the production of organic dyes. Government, academic, and industrial cooperation and a shared long-term vision helped Germany leap ahead in the invention and production of synthetic dyes. British and French dye manufacturers did not enjoy such favorable conditions and faced resistance to the new chemical dyes as a result of vested interest in the indigo and madder trades.[95]

By the late 1870s Germany had overtaken both Britain and France as the main producer of aniline and azo dyes. A close circle of German academic and industrial organic chemists dominated the knowledge making, education, and politics of organic chemistry and its applications. Germany and German chemists had a vested interest in promoting chemical dyes as an important new resource in an increasingly industrialized international marketplace.

Historians have argued that cultural differences in scientific education and training, government and institutional support, management techniques, corporate structure, and financing and patent legislation, as well as sophisticated sales and marketing, all contributed to Germany overtaking Britain and France in synthetic dye production in the later nineteenth century.[96] Any account also needs to consider geopolitics, including colonial empires and international trading patterns as well as networks of chemists and movements of knowledge.

By the end of the nineteenth century the commercial exploitation of coal tar chemistry, precipitated by the discovery of chemical dyes, was changing international trade and power relations. Britain and France could no longer rely on their far-flung colonies and pre-1850

Table 2.1 German chemical associations

1867	Deutsche Chemische Gesellschaft (German Chemical Society) (DGC) Journal: *Berichte der deutschen chemische Gesellschaft*	Founded and led by Hofmann to allow theoretical and industrial chemists to exchange ideas and information. Chaired by academic chemist Adolf Baeyer with statutes drawn up by industrial chemist Carl Martius. Emil Fischer succeeded Hofmann as director. The DCG became dominated by academic chemists.
1877	Verein zur Wahrung der Interessen der chemischen Industrie Deutschlands (Association for Protecting the Interests of the Chemical Industry of Germany)	100 chemical factory owners; leading industrialists; overlap with DCG.
1877	Verein Analytischer Chemiker (Association of Analytical Chemists)	Originally an association for professionals and chemists working in private and public testing laboratories.
1887	Became Deutsche Gesellschaft fur angewandte Chemie (German Society for Applied Chemistry) (DGAC) Journal: *Zeitschrift fur angewandte Chemie*	Expanded to include chemists working in other sectors.
1896	Became Verein Deutscher Chemiker (German Society for Applied Chemistry) (VDC)	Carl Duisberg, director of Bayer, reorganized DGAC to VDC to be the principal association for all German chemists. By 1900, senior VDC membership was dominated by factory owners and senior academic chemists.
1883	Freie Vereinigung Bayerischer Vertreter der angewandten Chemie (Free Association of Bavarian Representatives of Applied Chemistry)	
1901	Became Freie Vereinigung Deutscher Nahrungsmittelchemiker (Free Association of German Food Chemists) Journal: *Zeitschrift für Untersuchung der Nahrungs- und Genussmittel sowie Gebrauchsgegenstande (ZUNG)*	

industrial successes to maintain their status and dominance in world trade, as Germany built its own empire based on homegrown chemical foundations.

The chemists manufacturing chemical dyes were rapidly reconfiguring manufacturing, geopolitics, and consumer culture. Within a few years, food manufacturers and retailers across Europe and in America also appropriated the new dyes and recruited chemists as consultants to help to mediate the use of the dyes in food. As will be seen, many of the chemists involved in the mediation process were either German born or German educated or had links with German chemical companies.

Table 2.1 (*continued*)

1896	Verbrand selbstständiger öffentlicher Chemiker (Association of Independent Public Chemists) Journal: *Zeitschrift für offentliche Chemie*	
1897	Verband der Laboratoriumverstände an Deutschen Hochschulen (Association of German University and College Laboratory Directors)	
1900	Verein weiblicher Chemiker (Association of Women Chemists)	The VDC (above) did not allow female members until 1910.
1902	Deutsche Bunsen-Gesellschaft für angewandte physikalische Chemie (German Bunsen Society for Applied Physical Chemistry)	
1904	Bund der technischen-industriellen Beamten (League of Technical and Industrial Officials) (BTIB)	Set up by relatively low-paid and young German industrial chemists, demanding more rights and better working conditions and pay.
1905	Anschuss zur Wahrung der gemeinsamen Interessen des Chemikerstandes (Committee to Protect the Common interests of the Chemical Profession)	Formed by Carl Duisberg, partly as a response to the formation of the BTIB above, to promote the common interests of the chemical profession, including industrial employers. VDC cooperated with the public and food chemists to raise the profile and position of all chemists working in Germany.

Compiled from information in Johnson, "Germany: Discipline-Industry-Profession"; Cocks and Jarausch, *German Professions, 1800–1950*; Deutsche Chemische Gesellschaft, *Berichte der Deutschen Chemischen Gesellschaft*; Lepsius, *Festschrift zur Feier des 50 jährigen Bestehens der Deutschen Chemischen Gesellschaft*; Ruske, *100 Jahre Deutsche Chemische Gesellschaft*; Meinel and Scholz, *Die Allianz von Wissenschaft und Industrie*; Rassow, *Geschichte des Vereins Deutsche Chemiker in dem ersten 25 Jahren*; Burchardt, "Die Zusammernarbeit zwischen chemischer Industrie, Hochschulchemie und chemischen Verbänden in Wilhelmischen Deutschland," 194.

3

From dye manufacturer to food manufacturer

Although sold primarily to the textile industry, the new chemical dyes were used to color food almost as soon as they began to be manufactured commercially. Aniline dyes, and later azo dyes, often proved cheaper and more effective than alternative plant- and animal-based dyes such as saffron and cochineal. A tiny amount of the new substances could dye large quantities of food, and their excellent stability compared with many plant dyes made them more reliable in food processing procedures such as heating, mixing, and preserving. Manufacturers quickly began to exploit the extensive new range of colors to replace color lost after processing and to make manufactured food look more natural. However, the ability to reinforce color artificially also made it easier for unscrupulous food manufacturers to reduce costs through dilution or bulking out with cheaper ingredients, such as water added to milk or flour added to mustard. The new dyes also helped to extend the shelf life of products, as many of the colors worked as a preservative, having antioxidant properties.

From the beginning of their industrial production, dye manufacturers established close links with the textile industry for which the dyes were intended, using sophisticated marketing techniques and dedicated sales teams to advise and train textile manufacturers how to use the new dyes.[1] However, most dye manufacturers did not sell their dyes directly to the food industry. The use of the new dyes in food was never intended at the outset by dye manufacturers and, as the market for textile dyes was so much greater than that for food dyes, no attempts were made to understand how, or even whether, dyes could be used in food manufacturing. While most chemical companies simply ignored the use of their dyes in food, a few manufacturers actively discouraged it, as pointed out in a letter to *The Times* in 1884:

> There is every reason to fear that in these "cheap and nasty" times aniline dyes are being used to a considerable extent in the manufacture of sweets and confectionery, so much so indeed, that (to their honour be it said) the largest manufacturers of aniline in this country distinctly refuse to quote to confectioners: but there is nothing to prevent the unscrupulous maker of sweets from procuring such dyes from other parties or through third parties.[2]

Some dyemakers, though, were aware that their products were being used by retail chemists in drugs and tonics and sold as colorings and encouraged such use. As early as 1865, the Berlin-based aniline manufacturer G and A Levinstein, a German family dyemaking firm which subsequently set up in Manchester, England, announced that it was able to sell aniline dyes "free from any trace of arsenic, so that the aniline dyes can now safely be used by chemists and druggists."[3] Direct advertising and marketing to the food industry, however, was left to the retail and wholesale chemists and not carried out by the chemical manufacturers.

BASF's archivist has said "there are only very few indications in the corporate archives that BASF produced and/or sold food dyes in the nineteenth century," citing only a 1886/87 price-list containing several "dyes for confectioners."[4] However, while BASF did not market their dyes as food dyes or aggressively sell their dyes directly to the food industry, their textile dyes rapidly became extensively used as food colorings. Interestingly, in 1907, when the US government proposed certifying specific dyes that could be used for food, BASF "imperiously demanded" the inclusion of all their colors, and a simple form that they could fill out once to permanently register their colors as food colorings, according to Bernhard C. Hesse, the chemist commissioned to investigate the safety of food dyes. BASF's US representatives objected when told that each batch of dye might require separate tests and a guarantee of harmlessness, claiming that the company, like other dye manufacturers, did not test whether or not dyes sold for use in food were free of harmless contaminants, and indeed knew of no such tests.[5] Another dye manufacturer told Hesse that it never distinguished between textile dyes and food dyes, stating: "Napthol yellow S for wool and napthol yellow S for foods all comes out of the same barrel."[6] Following the introduction in the US of certification requirements for food dyes, BASF and most other German manufacturers withdrew from selling food dyes in the American market, opting instead to supply their textile dyes to US dye companies, who then purified them and obtained the certificates needed to guarantee that the dyes were safe to be used in food.[7]

Manufacturers play down concerns of regulators and the press

By the 1870s, the new chemical dyes were being used to color a wide range of food and drink products, passing through a complex international supply chain in which dyes were given new innocuous sounding names, such as "butter-yellow," that concealed their chemical origins.

It is significant that there is little mention in the press before 1870 of the substitution of mineral colors in food with the new aniline dyes being prepared for the textile industry. However, although there were fewer comments in the press about inappropriate colorings being used in food after the use of mineral dyes declined, there is little doubt that dyes were still being used to color food. Indeed, cookbooks at this time promoted "spectacle," highly colored food where presentation was deemed more important than taste.[8] Advertisements from wholesalers of food colors and essences discussed new "harmless" substances that could be used in minuscule amounts to color food safely, but there was no mention of what these substances are and from what they are made.

While *The Ladies Treasury* wrote, as early as 1865, about "the delicate flavouring of many confections, of custards, and of corn flour puddings" coming from coal tar, and other popular magazines referred to coal tar–derived flavorings, there was little mention in the press for at least a decade of aniline dyes being used to color food.[9] While the adulteration of food with cheap substitute ingredients such as flour in mustard, water in milk, and alum in bread was a growing political and social issue, there was at this time no association made in the media between these new artificial additives and the extensive public debate on adulteration. Indeed, most press comment alluded to science's ability to improve the aesthetic quality, availability, and accessibility of food products.

For the first few decades of their existence, aniline and other coal tar–derived dyes, manufactured specifically for the textile industry, were being incorporated into food, with few questions being asked of their suitability by food producers, retailers, ingredient wholesalers, public analysts, physicians, the public, politicians, or the press, despite the controversy surrounding poisonous metallic food colorings that raged for most of the nineteenth century. It seems, therefore, that these new substances were being introduced into the food supply with no publicity, no concern, and no monitoring, an example of Whorton's concept of new chemicals being freely and unsuspectingly welcomed into the domestic environment. The very same dyes that were being criticized in the press for harming the feet and constitution of colored sock wearers were being introduced into the public food supply with no adverse press coverage.

That aniline dyes were being used in food is evidenced by references in newspapers and periodicals such as *The Bradford Observer*, which in 1869 listed the wide range of uses for aniline dyes including inks, painting, photographs, "the soaking of tissues of objects for microscopical and anatomical purposes . . . the tinting of . . . white vinegar and syrup of raspberry, the blueing of linen and the colouring of confectionery." The article also points out the use of aniline dyes in cosmetics, which it claims "has been of undoubted service. They have superceded the metallic substances—the preparations of mercury, of bismuth and of lead—which were almost all injurious to health."[10] It seems, therefore, that the aniline dyes initially became a commonplace, unchallenged substitution for the tarnished and toxic metallic colors.

The *Bradford Observer* article was based on a review of Reimann's 1868 book *On Aniline and Its Derivatives*, as well as on Dr. Hofmann's *Report of the Colouring Matters Derived from Coal Tar Shown at the French Exhibition in 1867*. As in the other early reports on aniline dyes, the article was still very much in tune with the public rhetoric of the chemists. Probably a syndicated article that found its way into many newspapers around the country, it quoted *The Spectator*, writing that in 1856 "a young English chemist . . . discovered a new dye of unexampled brilliancy and beauty," observing that "no sooner was Mr Perkin's discovery made known than it riveted the attention of chemists in every country. A whole army of experimenters in England, France and Germany immediately commenced to prosecute further investigations." The article noted "the unprecedented rapidity with which the new dyes were adopted with the civilized nations of the west," and remarked that Europe "soon exported them to China, Japan and America . . . to those favoured climes which up to the present time had supplied the manufacturers of Europe with tinctorial products." Describing the synthetic dye industry as a "veritable revolution," the newspaper claimed that "chemistry, victorious, dispossessed the sun of a monopoly which it has hitherto always enjoyed."[11]

It is perhaps not surprising, then, that one of the earliest expressions of concern about aniline dyes in food is in a publication based in one of those countries that imported Europe's scientific "wonder product." An 1871 article in *The Health Reformer*, published by the American Health Reform Institute, titled "Poisoned Candies," warned that "the candy makers" of New York are spreading death among children since "various cheap devices are employed as substitutes for cochineal and saffron." According to the journal, the "red colour is usually produced by amboline, which is obtained in a crystallized form from coal tar during its process of refining." Sold for $2 an ounce, amboline "will equal in colouring twenty times its weight in cochineal." Other red dyes used included another aniline color, fuchsine.[12] This article suggested that aniline dyes were already making their way into food in the US by the early 1870s. However, there was remarkably little critical mention of their use in consumables in the British press at this time. Hassal acknowledged the widespread practice of coloring food with coal tar dyes in his influential *Food: Its Adulterations and the Methods for Their Detection* in 1876 but saw no reason to object to their use, as long as they were not contaminated with arsenic.[13]

Moreover, when critiques were aired, there was confusion as to whether the issue of concern should be the old antagonist arsenic or the new aniline substances. Arsenic became central to discussions about the possible toxicity of aniline dyes in food. This can be seen in an 1878 article in *the Analyst* describing a letter to *The Times* calling for action to be taken over the "poisonous nature of the colours used" in ice cream. The article pointed out that magenta, an aniline dye that sometimes contained arsenic, was used in ices. However, the article argued that it was "much more probable that any injurious effects which have been produced are due either to the indiscriminate use of ices by children when heated by over exertion, or surfeited with a quantity of indigestible food, or to the use of decaying fruit in the making of ices."[14] This article suggests that analysts were treating any claims as to the poisonous nature of the coal tar dyes with considerable circumspection, pointing to more "human" interventions, including contamination (by arsenic), the use of rotten fruit, and even bad parenting. It is perhaps not surprising that chemical analysts were reluctant to blame the use of chemicals, synthesized by chemists, for causing harm. In many ways the

rising respect for, and status of, chemists during this period was linked
to the dyes and other useful chemical products that chemists had managed to synthesize from coal tar.

Foreign poison

However, from the late 1870s onwards, there is evidence of mounting
concern in the media with the identification of the synthetic food colorants as a new form of adulteration, rather than a safe replacement for
the poisonous substances previously used to color food such as copper,
lead, and arsenic. It is hard to pinpoint exactly why the use of synthetic
dyes in food began to be questioned. It may well be linked to the rise of
the French, German, and Swiss chemical industries at this time and the
increase in imported food, such as the newly patented oleomargarine, a
cheap substitute for butter, which was often dyed yellow to make it appear more like butter.[15] Certainly, one of the noticeable trends in reporting from this time was the association made between these new food
adulterants and foreign imported food. As historians of food regulation
are beginning to show, regulation is often used as an instrument to limit
food imports, and reportage in the popular press shows a propensity
to blame imported food and foreign chemists for adulteration scares.
An 1878 edition of *Funny Folks* described "the efforts made by an in-
genious French chemist to give us our mutton and beef and pork by a
dozen different flavours." The journalist added:

> A German chemist is now experimenting with the view of giv-
> ing us our mutton, beef and pork of as many different colours.
> The eye, he thinks, is weary of the monotonous tints of fat and
> lean and why, he would seem to be anxious to know, should
> not a mutton-chop of azure, with magenta fat, let us say, be a
> possibility? Why should not a purple haunch of mutton be kept
> in countenance by an emerald-green round of beef, flanked by
> cutlets of a vivid orange, and sweetbread of a deep blood-red?[16]

By the 1880s, tinted foreign produce was under sustained attack.
The Country Gentleman warned that cheap French red wine full of aniline dyes was the cause of ill health among the British population, while
the whimsically titled *Moonshine* pointed out in 1886 that in Paris
"they make quite a nice *genuine* raspberry jelly" made from "oil of vit-
riol, impure glucose, algine, aniline and raspberry ether."[17] In 1890 *The
Sheffield and Rotherham Independent* mentioned a report on Bologna

sausages that are colored with "garnet red and Bismarck red." It was stated, the newspaper noted, "that purchasers 'like a bit of colour' but one cannot but suspect that these aniline dyes are used to prevent them knowing what they are eating."[18] Meanwhile, according to *The Friendly Companion: A Magazine for Youth*, orange growers in Florida "manufacture" blood oranges by "piercing the rind of ordinary oranges with a fine syringe and injecting an aniline dye which quickly permeates the pulp."[19] The widespread criticism of foreign food, and claims that it was adulterated and possibly poisoned, were tied up with prevailing concerns over food shortages for an urbanizing population. In an effort to avoid shortage and increase variety, more food was being imported, yet its origins and manufacture were increasingly inscrutable. The deepening international hostility and rivalry of the 1870s, both economic and scientific, also contributed to such attacks.

By 1890, it appeared as if science, far from producing safe new substances to replace the harmful poisons of the past, was simply adding to an earlier list of food adulterants available to deceive and possibly harm an unwitting public. This was illustrated amusingly in a "modern fable" told in an August 1890 edition of *Illustrated Chips* about four flies. The first fly "applied himself to a dish of milk, but ere long died in convulsions, produced by the chalk that was in the milk. The second found a sausage . . . but the sausage having been coloured with aniline, the creature was presently poisoned. A like fate befell the third, who had eaten flour containing . . . alum." The final fly, "filled with despair," attempted suicide by sucking up the liquid on some flypaper but, instead of dying, found himself refreshed and vigorous. "Even the fly-paper had been adulterated."[20] From here on in, the Victorian media presented aniline dyes as adulterants. British food and drug manufacturers began to exploit the growing public unease by advertising their products as free from artificial food colorings, or, in the case of Eno's Salts, a relief from "the great danger of poisonous aniline dyes in sugar, and pink or chemically-coloured sherbet."[21]

This change in the public's view of food coloring coincided with a wider set of programs, such as the arts and crafts movement, which explicitly rejected industrial goods in favor of a return to natural products. Food historians have observed that from the 1890s, vividly colored food, formerly highly esteemed in Victorian grand cuisine, had come to be seen as an artifice and a form of social climbing.[22]

Within four decades of their invention, synthetically manufactured chemical dyes had thus been transformed by both the general and specialist press from a wonderful new substance of science into a toxic tool for commercial deception, an out-of-control danger in need of correctives. Many groups of people, including manufacturers, retailers, professional scientists, and politicians, used the media to further their own interests. However, it is also clear that the press took a considered and critical approach to both chemistry and commerce. While news reporting to some degree reflected public opinion and the views of other interested parties, how issues were reported in the media also helped form public perception, making the information flow a two-way process.

Interestingly, during this period of growing awareness and doubt expressed in the press, it is still very unclear from articles or advertisements in the popular press who was actually employing aniline dyes in food. While manufacturers were increasingly likely to advertise their food as "pure" and free from adulterants during the nineteenth century, evidence from company archives and parliamentary investigations suggests that aniline and other coal tar–derived dyes were in fact being extensively used in food preparation, despite nobody laying claim to such practices. Some historians have suggested that as science became increasingly specialized and laboratory based during the 1870s, it became more detached from the wider public.[23] However, the coverage of the new coal tar dyes suggests that the popular, professional, and trade press, particularly in the latter years of the nineteenth century, was far from a passive organ for relaying the progress of science. Creating a forum for debate and an ability to reach an ever-widening audience of readers with often insightful, irreverent, questioning, and critical commentary on science and its application, the press was much more than a simple disseminator of science.

By studying the trajectory of these new synthetic substances through the expanding Victorian media over several decades, we can see how they were transformed from their original profiles, determined by scientists, into something with a different presence and significance, mediated by a wide range of readers and writers, both lay and scientific. The changed perception of aniline dyes and their use in food led to transformations in both the substances themselves and their use. The press-mediated image of dyes in food in the nineteenth century also helped to transform scientific careers, producing new roles, such as the consultant scientist, the public analyst, and the food chemist, as well as encouraging new ways of controlling such substances through regulations and the creation of government, private, and consumer-based institutions and organizations.

Manufacturers defend the use of chemical dyes in food

When the extensive use of aniline and azo dyes in food and drink production became a public concern during the 1880s, dye manufacturers
responded defensively to media discussion of the subject. Both chemical
companies and food manufacturers either denied knowledge of the use of
the dyes in food production, or argued that they were harmless, especially
in the small quantities used. Both groups also began to use scientists to
verify their claims. At the request of BASF, Hofmann and the hygienist
Rudolf Virchow publicly vouched for the harmlessness of many aniline
and azo dyes produced by the company. While Hofmann was one of
Germany's most respected chemists, Virchow was equally well known as
one of the country's leading medical scientists working on cell and cancer
theories. Virchow was also a prominent politician and public health campaigner, whose investigations into diseases caused by roundworm found
in pork led to the introduction of meat inspection in Berlin.[24]

The apparent endorsement of aniline dye use in food by these prominent German scientists was acknowledged in an 1887 publication by
Paul Cazeneuve, professor of chemistry and toxicology at the Lyon
medical faculty, and published jointly with BASF's factory in Neuvillesur-Saône, Lyon. The publication included experimental and scientific
evidence, produced by Cazeneuve and other physiological and organic
chemists, purporting to show the harmlessness of aniline and azo dyes.
As well as reproducing the proclamations signed by Hofmann and Virchow, the publication also included statements by Carl Gräbe, at the
time professor of chemistry at Geneva University and a consultant to
BASF's French factory, and other notable chemists, guaranteeing the
harmlessness of most chemical dyes.[25]

It is no coincidence that BASF commissioned a publication on the
safety of aniline and azo dyes, endorsed by high-profile chemists, in
1887. At precisely this juncture, governments in Germany and France
were reviewing food legislation amid media concerns across Europe
about the possible toxicity of the new chemical dyes. As table 3.1 shows,
legislation on the use of the new dyes for food coloring began to be
implemented in varying degrees across Europe and in the US, although
notably not in Britain.

Several nations had banned specific poisonous mineral colors earlier
in the century and subsequently banned some of the new chemical dyes,
particularly those processed by the arsenical method. However, efforts
by chemists and the dye industry to prove chemical dyes were safe kept
dozens of the new dyes from being banned by European legislators. For

Table 3.1 Legislation relating to the use of synthetic dyes using the classification of dyes as described in the 1908 Green Tables, a classification system named after Arthur G. Green.

Austria	Law of March 1, 1886, banned No. 483 of the Green Tables and all aniline colors.
	Law of May 1, 1886, banned No.1 of the Green Tables and all aniline colors.
	Law of September 19, 1895, permitted just 16 types of dye and banned all others. Effectively the law permitted 47 entries in the Green Tables.
	Law of January 22, 1896 permitted 22 dyes listed in the Green Tables.
Belgium	Law of 1891 banned four dyes listed in the Green Tables.
France	Police ordinance of May 21, 1885, banned 489 dyes listed in the Green Tables.
	Police ordinance of December 31, 1890, banned 469 dyes listed in the Green Tables, reinstating 23 Green Table dyes that had previously been banned in the 1885 ordinance.
Germany	Law of July 5, 1887, specifically banned Nos. 1 and 483 of the Green Tables. The general interpretation of this law effectively meant that all other Green Table dyes were considered to be permitted for use in foods.
Italy	Law of February 7, 1892, banned all dyes except for 9 types of dye, effectively permitting 32 dyes listed in the Green Tables.
	Law of February 7, 1902, specifically banned 37 entries in the Green Tables and permitted 11 specific dyes.
	Decree of June 29, 1893, permitted 7 types of colors with definitions sufficiently elastic to include 34 different dyes listed by Green.
	Decree of March 24, 1895, banned 4 dyes listed in the Green Tables.
US	1907 Food Inspection Decision No. 76 permitted 7 dyes that were required to have a guarantee from the manufacturer stating that they were free from subsidiary products and may be used in food.

Sources: Hesse, *Coal-Tar Colors Used in Food Products*, 5; Green, Schultz, and Julius, *A Systematic Survey of the Organic Colouring Matters*. For more on the Green classification, see chap. 4.

example, the German law of 1887 distinguished between contaminated chemical dyes and pure aniline and azo dyes, while the Paris Prefecture of Police removed a ban on fuchsine in confectioners' produce in 1890.[26] These laws had a twofold effect. Firstly, they led to the prohibiting of certain dyes, including, for example, the banning in Germany of all colorings containing arsenic, copper, zinc, tin, corallin, picric acid, and other named substances. This provided an incentive to find alternative colorings. Secondly, lists of prohibited dyes suggested that any dyes made from substances not on the list could be legitimately used in food and drink production. As a result, wholesale and retail chemists were able to claim that many of the aniline and azo dyes they were selling were "harmless" and in accordance with food legislation. These chemists advertised the new "harmless" dyes widely in journals aimed at the food industry and domestic consumers. Even a few chemical manufacturers decided to turn their attention to the food sector at this point, as a means of attracting new business.

One company that opted to specialize in supplying the food sector with the new dyes was the Berlin-based dye company B R Brückner Nahrungsmittelfarben-Fabrik, which claimed in advertisements and flyers that its dyes were "giftfrei" (nontoxic) dyes for food and drink, as they were made in accordance with the 1887 German food regulations. Brückner boasted that their dyes made all meat look natural, requiring only fifteen grams of dye for forty-five pounds of meat. Indeed, one advertisement claimed that the "grey, slimy and untrustworthy" characteristics of poorly processed and lightly smoked sausage meat "can be easily and thoroughly remedied with Brückner's harmless dyes." The dyes, used to "beautify and preserve sausages," were sometimes sold with an accompanying certificate allowing food producers to claim that they were acting in accordance with legislation. One such certificate stated that "I declare to the consumer that I have used 20 gm of Brückner's concentrated sausage coloring (*Wursttinctur*) or powder (*Carminsurrogat*) to each 1 kg of salt to stop the greying of these sausage products. These surrogate dyes comply with the [law of] 5 July 1887 as not injurious to health."[27]

However, because the food sector remained a minor consumer of the new dyes compared with Europe's flourishing textile industry, most dye manufacturers continued to focus their marketing and production on textile dyes. As a result, virtually all the dyes used by food producers until well into the twentieth century were essentially textile dyes, sold by wholesale and retail chemists under a wide range of attractive names devised by the retailers themselves which provided their purchasers with no clues as to their chemical origins.

Among the earliest dye manufacturers to produce dyes specifically aimed at the food industry were the US dye manufacturers H. Kohnstamm & Co. of New York and Schoellkopf, Hartford and Hanno of Buffalo, NY which manufactured and sold food dyes through its National Aniline Co. It is noteworthy how many US-based chemists were German, had German ancestors, or had spent part of their chemical training in Germany. Kohnstamm and Schoellkopf marketed their dyes as "certified food dyes," after the US government introduced legislation which recommended that only seven specific coal tar dyes, out of hundreds of similar dyes readily available in the marketplace, could be safely used in food and drink.[28] These American dye manufacturers worked closely with the US government's Bureau of Chemistry to produce purer dyes specifically aimed at the food market. H. Kohnstamm imported German dyes before purifying them and then sending them to the bureau for inspection, while Schoelkopf and Hanno manufactured their own purified dyes specifically for the food industry. Indeed, Kohnstamm had been working on producing dyes specifically for use in food for at least two decades before the government intervention, employing the Hofmann-trained American chemist Hermann Endemann, editor of the *Journal of the American Chemical Society*, to develop suitable colors for food, which were approved by the National Confectioners' Association as early as 1884.

However, these dye companies were in the minority. Bernhard Hesse, a consultant chemist commissioned by the US Department of Agriculture to determine how best to regulate dyes for food, was shocked at the vast numbers of coal tar dyes being used as food dyes, very few of which had been tested or proved to be harmless. Hesse noted that even though the dyes were being widely used to color food and drink, most dye manufacturers showed no effort to work with the food industry or provide suitable dyes or guarantees, as they did with the textile industry. In his view, dye companies showed a lack of care and knowledge about how their dyes were being used, with one chemist informing him that concentrations of dye usage of 1 part in 3,000 were not unusual, while another recommended no more than 1 part in 50,000.[29] The lack of standardization in the use of such dyes remained an issue for many decades.

One of the earliest companies in Europe to produce dyes specifically for the food industry was London-based Williams Bros. As a result of the increasingly negative publicity surrounding the use of aniline dyes in food, Williams decided to select certain dyes that could be manufactured without arsenic and commissioned the consultant analytical chemist Samuel Rideal to test food colored by the dyes on animals. The company then marketed the dyes to the food trade as "Harmless Colours for Confectionery."[30]

Williams later explained how the enactment of separate legislation in different countries had created a niche market for dye companies choosing to specialize in marketing food dyes. The company had

> made a special feature of the manufacture of confectionery colours to meet the laws of all the leading countries of the world, and the shades in most demand have all been matched exactly and tabulated, so that a colour to satisfy the requirements of any particular country can be referred to and supplied promptly.

> The high standard of such colours, which have to conform to a
> very drastic schedule as regards freedom from all metallic im-
> purities, moisture, sediment etc has practically taken them out
> of the orbit of aniline dyes and raised them to the status of fine
> [i.e., specialist] chemicals.[31]

Williams had thus preempted legislation in an attempt to resolve problems of standardization, color identification, and testing that would pre-
occupy chemists in the public domain, as later chapters will show. Even
by 1936, the time of Williams's comments, however, there was still no
consensus as to which dyes could be considered safe and which should
be banned, and the company noted that "these special ranges of permitted dyes are not always by any means the most judiciously chosen by
the countries concerned, but as the authorities are particularly strict in
adhering to their own permitted colours, it is incumbent upon shippers
to see that no others are used."[32]

Food producers' lack of awareness and disclosure

For many years, food producers failed to acknowledge that chemical
colorings were being added to food and drink, partly because they did
not know, or chose not to investigate, what the dyes they were using
were made from. The extent to which food manufacturers were aware
of the nature of the dyes being used as food colorings was made clear
in their responses in 1901 to a Parliamentary Select Committee set up
by the British government to investigate the use of chemical additives in
food. When Robert McCracken, managing director of United Creameries Ltd., was asked if he knew what aniline color his dairy used in the
manufacture of butter or margarine, he replied, "I really do not know
what the actual article is." His answers provide a revealing insight into
the ignorance of the food manufacturers about the dyes they were using
and the lack of transparency throughout the supply chain.

> *You do not know what they are called?*—No; I find that there
> has been aniline colouring in one of the butter colours which we
> have used. . . . but the quantity is so small that it would practi-
> cally do no harm.
>
> *Where do you get these aniline colouring matters from?*—They
> are mostly from Continental manufacturers.

You do not know the chemical names of them or the trade names of them?—No.

Nor anything about them?—No, we buy them as butter colouring . . . and we get a solution in oil just as we get annatto.

Only you know the difference that one is annatto and one is an aniline colour?—We have been told so.

Is it on the bottle?—No, but the manufacturer informs us that there is a proportion of aniline colour along with annatto in it.

Is there a declaration on the bottles of all these different colouring matters as to what they consist of?—No.

How do you know the difference between them?—The manufacturer has told us that he does use a little aniline colouring in making up the colour.[33]

McCracken's responses demonstrated substantial trust in the suppliers of dyes, as well as a lack of concern on the part of the food manufacturer about the actual substances used. He added that his company never tested the dyes because "we expect that we are dealing with honest people, and that they give us what we ask for." He told the committee that he thought the aniline dyes he used were Dutch but admitted to having no knowledge as to their nature and to being unaware of the use of Martius yellow as a butter coloring. When asked if some of the aniline dyes used to color margarine and butter "really are most noxious substances" he simply answered: "they may be."[34]

The testimony of Leonard Kidgell Boseley, a chemical analyst employed by the preserve and confectionery manufacturer James Keiller and Son Ltd., based in Dundee, Scotland, displayed a more informed knowledge of the coal tar dyes used in food manufacturing.[35] Boseley had previously worked with one of Britain's foremost public analysts, the German-born Otto Hehner, and had been an analyst with the Aylesbury Dairy Co. He provided the committee with detailed tables of the different types of aniline dyes used for different foods and the quantities typically applied. He also referred to experiments of colorings on digestion by H. A. Weber.[36] However, even this trained chemist demonstrated a high degree of trust in his suppliers and a lack of interest in investigating the dyes himself.

According to Boseley, the coloring matter used in dark-red jams "is magenta or fuchsine, which is an acetate of rosaniline, or a hydrochlorate of rosanilin," with about half a gram of dye added to a hundredweight of jam. However, "there is another colouring matter, which is called vermilion, and is a lighter red, which is also used by jam manufacturers. I am sorry I cannot give you the actual composition of that, but I know it is an aniline dye," he added.[37]

After the abolition of sugar duty in Britain in 1874, manufactured jam had become a more affordable product for lower-income families, with bread and jam becoming a staple diet for poor children. By the end of the nineteenth century, the jam market was highly competitive, and manufacturers of jam intended for the lower-income consumers resorted to using artificial dyes to reduce costs.[38] When questioned about the possible health effects of the dyes Boseley told the committee that Keiller never used metallic dyes and that the aniline dyes it used, mainly in confectionery products, were "about as free from arsenic and lead as they can possibly be." Boseley's comments show that, even as late as 1901, analysts still were focusing on the possible metallic contamination of chemical dyes, rather than the possible toxicity of the chemical substances themselves. When asked if he had tested the dyes himself, Boseley replied that he had not and appeared to put total trust in his supplier, the London-based dye trader Harker, Stagg & Morgan, who sourced magenta dyes from Germany. Boseley told the committee that he understood from Harker's chemist, "a man I know perfectly well, that the people who mix these colours, and work with the colouring matters inhale a great deal of it in the form of dust and so on into their lungs, and their health never seems to be impaired. I mean people who have worked there for six years. I rather believe that he guarantees several of these colours as being absolutely harmless." When pressed by Professor Thomas Thorpe, vice-president of the Royal Society, Boseley stressed that the dyes "are sold by Messrs. Harker, Stagg & Morgan as being absolutely harmless. In every case we have that put down," but admitted that he had "not verified that statement."[39] Even though some food manufacturers were employing chemists by the end of the nineteenth century, this testimony shows that knowledge of, and investigations into, aniline and azo dyes remained limited, despite ongoing concerns over their use in food.

Criticism from the sanitary press

However, concern about the chemicalization of everyday food products continued to grow and some of the most vociferous complaints in the press about the use of the new dyes in food came from a new type of campaigning social hygiene and antiadulteration publication. Several of these publications were launched during this period to campaign for "pure food" and greater food regulation.[40] One such publication was the weekly magazine *The Public Analytical Journal and Sanitary Review*, also known as *Food, Drugs and Drink*. The journal, launched in 1892, changed its name to *Food and Sanitation* in 1893 and claimed that its aim was to promote "the enforcement of the Food and Drugs Act, not in the interests of Public Analysts, or of any class or profession, but of the vendors of unadulterated articles, and of the consuming public."[41] The journal, which sought contributions from among the public health and analytical community as well as independent grocers, denounced chemical coloring as a new form of adulteration that was effectively being slowly legitimized by manufacturers, with no effective resistance from analysts.[42] While the anonymous author noted that most of the coloring used was "entirely harmless, and has no influence upon the health of the consumer," he or she expressed concern that several aniline colors, including Martius yellow, safranine, methylene blue, dinitro cresol, and aurantia, were known to be poisonous even in a pure form, while others became poisonous when contaminated through the manufacturing process with impurities such as arsenic, copper, tin, or zinc. Meanwhile, the author claimed, most manufacturers had no idea what sort of chemicals they were using, and that "even if the quantity of colour consumed by one individual may be exceedingly small, it is palpably evident that no such manufacturer should have the right to use such colours."[43]

The article's author expressed bemusement and concern as to why "almost every civilized country has passed legislative measures regulating the use of colours such as those referred to, but England, the mother of all Adulteration Acts, has done nothing, and has in this, as in many other points concerning the sale of food and drink, remained far behind most other nations, even including the squabbling South American republics." The points raised here are strikingly similar to comments made by the leading public analyst Otto Hehner in the Society of Public Analysts' journal the *Analyst* two years earlier, suggesting not only an awareness of the analysts' debates but the possibility that some of the anonymous contributions to *Food and Sanitation* may even have come from Hehner himself.[44] According to the article, the practice had continued unchecked because consumers had become used to artificially dyed food, but the author noted that this was no reason for "an evil habit, based originally and mainly upon fraud" to be allowed to continue.

> Let the private consumer use whatever colour he likes. Let him
> use cochineal for his jellies and ices, or saffron for his pie-crusts
> as our grandmothers used to do, but deliver us from the ama-
> teur dyer, who insists that almost every article of food which
> we have to purchase is dyed with colours of which he himself
> knows nothing, and of which the unfortunate consumer knows
> even less.[45]

This clearly demonstrates the issues at stake in arbitrating between
consumer liberty, self-determination and expectation, freedom of trade,
greater transparency and knowledge of commercial practices, state control, and consumer protection.

The article called for legislation specifically setting limits on the use
of dyes, defining which colors could be used or, better still, prohibiting
the use of all artificial coloring for food.[46] By the early 1890s several
European countries had begun to ban many of the new dyes, while in
1907 the US government recommended just seven named coal tar dyes
for use in food and drink. But despite ongoing criticism of the new dyes
in the Victorian press, no legislation for coal tar dyes was adopted in
Britain until 1925.[47]

The relationship between the "pure food" and sanitation magazines,
the analytical and health professions and the food industry is a complex
one. Public analysts contributed to several publications, and one leading
analyst, Charles Cassal, actually set up his own journal to campaign for
greater food purity and regulations. *The British Food Journal and Analytical Review* was launched in 1899 by Cassal and his mentor, the hygienist William Henry Corfield, professor of hygiene and public health at
University College London; it was edited by Cassal until 1914.[48] Other
campaigning journals sought to distance themselves from both the food
regulators and the food industry. *Food and Sanitation* (and formerly as
Food, Drugs and Drink) frequently denied that it was a mouthpiece for
medical health officers or public analysts—"we hold no brief for the
public analysts, neither officially nor semi-officially, nor have we any
connection whatever with them"—and indeed, the journal often criticized analysts for their inaction regarding, for example, the use of chemical dyes.[49] However, *Food and Sanitation* depended on the public health
profession both for its readership and its content, and often editorially
fought the same battles as the public analysts.[50] The journal also sought
to expand its readership among independent grocers, and campaigned
for a reform of the Adulteration Acts to ensure that both food producers
and wholesalers bore responsibility for consequences for adulteration

similar to that borne by retailers. According to the journal, "the wholesale dealer or manufacturer who practises fraud on a colossal scale" was
the real person whom the acts should reach, noting however, that

> the maker of adulterated mustard, cocoa, butter, or lard, has
> one great advantage over the retailer—he sits often in parlia-
> ment. . . . Grocers ought not to forget that the Food and Drugs
> Acts were passed by a House of Commons that did not contain
> one retailer, but was composed of wholesale manufacturers,
> brewers and other interested persons by the score, many of them
> reaping fortunes by fraud, and therefore having every interest in
> the maintenance of it.[51]

While public campaigners and even chemists themselves raised concerns about the proliferation of synthesized chemicals being added to
food, manufacturers of food were increasingly employing chemists as
consultants to help legitimize the use of the new substances.

Transforming food using chemists and colors

The food industry itself was in the midst of an industrial transformation in the later nineteenth century, which altered both the production
and distribution of many food items, from basic commodities such as
bread and milk to completely new food items, such as custard, baking
powder, and canned meat. As hundreds of thousands of people moved
into cities to work, the need to store and transport food increased, and
methods of preserving food improved. Dramatic structural change resulted in an increase of branded foods, the emergence of large multiple
retailers, and the growth of large food manufacturers and importers.
These included many companies that are still prominent today, such as
Nestlé in Switzerland, Crosse and Blackwell and Sainsbury in Britain,
Heinz in the US, and the widespread cooperative movement. By the end
of the nineteenth century, large food producers, retailers, and distributors had become powerful players in an increasingly industrialized and
international food sector.[52] Recent histories of sugar and baking powder, for example, demonstrate how politically and financially influential
industrial food companies became during this period.[53]

The sugar industry was one of the first food industries to employ
chemists as both analysts and engineers. During the nineteenth century,
sugar chemists in Europe and the US produced sugar from an increasing
variety of sources, including beet, sorghum, and corn as well as cane. As

demand for sugar soared and governments, particularly the US, began to impose tax on sugar imports, chemists were also employed to test for the purity of sugar using polariscopes and color tests.[54]

Chemists played an important role in the transformation of the food sector, both in devising new foods and production and preservation techniques and in the creation of expert systems of food surveillance. As firms sought to consolidate their position and power in the marketplace, they increasingly turned to chemists and chemistry to standardize processes and products, to prevent fraudulent adulteration or contamination in the supply chain, and to endorse their products as "pure" and "hygienic." Chemists, and chemical preservatives and colorings, became part of the armory employed by the large food production and distribution companies in their drive to secure market share and to ensure that their products were seen as trustworthy, reliable, consistent, unadulterated, and of good value.[55]

However, while a few chemists, such as Britain's Alfred Bird, the American Eben Horsford, and Germany's August Oetker, helped to create food companies on the back of chemical developments, such as the formulation of baking powders, most chemists involved in the food industry in the nineteenth century, outside of the sugar industry, were employed principally as consultants, either by the state to help identify food adulteration, or by food producers and retailers to help in their defense against prosecution for adulterating food.[56]

The complex and contradictory claims surrounding what constituted an adulterant as opposed to a food improver, as well as the conflicting and multifaceted relationship between the food industry and chemists, are evident from the archives of nineteenth-century food manufacturers. Rowntree's archives, for example, contain correspondence between food manufacturers and consulting chemists from the 1870s as well as increasing transactions between wholesale chemists and chemical manufacturers like London-based W J Bush & Co., A. Boakes, Roberts and Co., and Ehrenfest & Co., all of whom manufactured foodcolorings, flavorings, and essences.[57] The evidence suggests that food manufacturers and chemists worked closely together to manage consumer concerns and expectations.

Letters between the York-based chocolate manufacturer H. I. Rowntree and Co. and various analysts, including Hassall, illustrate the reciprocal and self-serving nature of their relationship. The letters demonstrate how food companies and chemists worked together to legitimize the use of chemicals in food, as preservatives and colourings, but also to generally improve food production. For example, in one exchange, the chocolate company stated that it was prepared to pay for an analytical report on its chocolate product provided that the results could be used for advertising purposes, going so far as to suggest that Hassall alter the wording if necessary in order to promote the product better!

> Oct 2nd [188]8
> Letter from H. I. Rowntree to Dr. Arthur H. Hassall, London
>
> Dear Sir,
>
> The price you name 15.15.0 pounds is high, but we would not demur to the amount for a report of a chocolate so worded as to be useful for publication. We do not, of course, ask in any way to anticipate the results of your enquiry and we have absolute confidence in everything that we put forth on behalf of Elect Cocoa. Still it is possible for an approving report to be so worded as to be of but little use for trade purposes and we therefore should like to enquire whether details of wording, if not contrary to the results of your enquiry, could be modified by you, in accordance our suggestions, without extra charge.

> Oct 4 [188]8
> Letter from H. I. Rowntree to Dr. Arthur H. Hassall
>
> Dear Sir,
> We are in receipt of yours of yesterday and thank you for the satisfactory answer to our query. Enclosed please find cheque for 15.12.0 pounds. As we are just commencing an Advertising campaign up in the North we shall be very glad to have your report as early as possible.[58]

After receiving Hassall's report, further correspondence shows a disagreement between the analyst and the food manufacturer as to whether the addition of certain substances constitute adulteration or should be viewed as ingredients used to improve the product.

> 16 Oct [188]8
> Letter from H. I. Rowntree to Messrs. Hassall & Clayton
>
> 54 Holborn Viaduct EC
>
> Gentlemen, We are in receipt of your letter of the 12th October and accompanying report, which is satisfactory, and beg to thank you for your early attention to the matter.

We feel the force of what you say in the concluding paragraph of your letter. But so far as we know the acrid taste of the pure cocoa can only be overcome in one of two ways—either by the addition of foreign substances such as sugar or farine—or otherwise by the application of a *small* quantity of pure alkali in the process of manufacture. Are we not right in thinking that dietetically our Elect Extract is in no way damaged, but the reverse, by the acid in the cocoa being neutralised by this small percentage of alkali?[59]

Rowntree was attempting to persuade Hassall that "pure alkali," probably in the form of potash (a form of potassium carbonate), was a preferable ingredient to use to overcome the "acrid taste" of pure cocoa than the addition of sugar or flour. Adding potash or other alkalis was known as Dutching because of its use by some European, notably Dutch, chocolate manufacturers to both neutralize the acidity of the natural cocoa and also to impart a darker color to the chocolate.[60] However, the use of potash was contentious at a time when chocolate companies such as Fry's, Cadbury's, and Rowntree's were seeking to market their products as "pure" and "free from adulteration."

Correspondence in the Rowntree archives demonstrates the extent to which the company kept secret its use of potash and Fuller's earth, a type of clay used to emulsify cocoa. In letters marked "private," sent to various chemical manufacturers and wholesalers, Rowntree requested that the supplies of these undisclosed components of the chocolate manufacturing process be sent not to the factory but to other addresses in York and that all correspondence be marked as confidential. Clearly, Rowntree was concerned to ensure that its use of such additives was not disclosed publicly, both to maintain the integrity of its products as pure and unadulterated and to prevent other chocolate manufacturers copying its techniques and ingredients.[61]

Correspondence in the archives indicates that chemists were regularly consulted on the use and identification of different ingredients and processes to improve food products, as well as to help defend the company against accusations of adulteration. Other paid services provided by analysts included assessing the company's products in a form that could also be used to promote them in the media, as described earlier. The archives include "confidential" correspondence with chemists throughout Europe and as far away as New Zealand, offering them money to disclose the secret industrial processes and ingredients used by competitor chocolate manufacturers. This type of correspondence was a two-way affair; there is evidence of chemists, and their patent lawyers, approaching Rowntree with information on new techniques or ingredients to improve its products, as well as offering to sell information regarding the practices and ingredients used by competitor companies.[62]

This evidence demonstrates how slippery the issue of adulteration was and the extent to which chemists were caught in the middle of an ongoing debate as to which food manipulations and additives constituted adulteration and which were innovations leading to product improvement. This was an era when food manufacturers recognized that changing the constitution of products was an important device in improving profits and market share. Chemists became an integral part of the whole industrializing food process, acting as consultants to the food industry while at the same time policing the national food supplies as public analysts.

The use of the new colorings in food also was subject to the same type of analysis, debate, and subterfuge within the food industry as chemicals used as preservatives. On February 12, 1892, Rowntree wrote to Ehrenfest & Co. in Blackfriars, London, which described itself as a manufacturer of harmless food colorings, seeking details of the chemist's claims to be able to offer a formula for manufacturing cocoa by the Dutch method "producing a result equal to the best brands" as well as "an innocuous colouring for deepening cocoa" and a "powerful and agreeable flavouring for the same as used also by the Dutch manufacturers." In a letter the following day, Rowntree invited Mr. Ehrenfest to visit the factory.[63] In June 1893, Ehrenfest & Co. wrote to Rowntree promoting a new soluble cocoa coloring that could be "used in infinitesimal quantities—a mere speck or two to a quantity sufficient to fill a large cup" for the purpose of strengthening the color of Cocoa Essence or "imitating the depth of the Dutch makes." The chemist noted that

> there is hardly a house of any size in England who is now not using it [as] it is *perfectly harmless inexpensive in use* (because so little is required) and *its effect on the appearance of the cocoa is wonderful.* We hope you will at least *try the effect* of this soluble colouring even if you do not order any of it by mixing a cup of your Cocoa Essence *then another by its side with a speck or two of the colouring added*: you will see the extraordinary difference. Its use in cocoa is as innocent as that of *colouring in Jellies & sweets, Anatto in Butter etc etc.*[64]

Judging by the description of this coloring and the quantities used, the new dye being promoted by Ehrenfest was almost certainly a coal tar dye. By the 1890s, German manufacturing and wholesale chemists were promoting aniline and azo dyes as harmless, a position adopted by wholesale and retail chemists across Europe. As explained earlier, this practice was given credence by the German 1887 food legislation and similar laws in other European countries that had banned a few named coal tar dyes, creating the impression that dyes not named in the legislation were safe for consumption. The use of chemical colorings can be seen in the Rowntree recipe books for chocolate and fruit gums, where there are mentions of dyes including Bismarck brown and rhodamine. Due to the huge variety of names used to describe colorings sold by chemists, it is difficult to say for sure how many of the other colorings listed in the recipe books, such as custard yellow, were coal tar synthesized or plant-based dyes, but coal tar dyes were being extensively used in food preparation, particularly confectionery, throughout this period.[65]

Trade names for coal tar dyes at the time included heliotrope, lavender, and Damson blue, all variations of Hoffman's violet aniline; chocolate brown, an aniline color, used in chocolate buttons in ratios between 1 in 33,000 and 1 in 17,000; primrose yellow, or auramine; vermilion and rose red, both aniline colors; and saffron yellow, an aniline dye, among many others.[66] Other coal tar color names used in the food industry and identified by the government laboratory were Ponceau red and citron orange in temperance beverages (what we now call soft drinks); crocein orange, citron orange, auramin, rose pink, and fuchsin in fruit, jellies, and jams; Congo red, fuchsin, and various sulphonated azo-reds in sausages; a mixture of azo red and Bismarck brown used to imitate the smoke color of hams; and acid-yellow (mixed with Bismarck brown in some cases) for coloring sugar crystals. Public analysts providing evidence to the 1901 parliamentary committee investigating the use of colors and preservatives in food stated that they had identified diazo-colors (both yellow and red) in table jellies; eosin and tropaeolins in sweetmeats, Congo red in sausages, a color resembling benzo-purpurin in jam, eosin dye and red anilines in meat products, and, increasingly, coal tar yellows of the tropaeolin group in milk and dairy products.[67] Fanciful color names given to some of these dyes, such as Silver Churn and Cowslip Colouring sold to the milk industry, gave no indication as to the composition of the dyes.[68]

The private Rowntree notebooks reveal that the company expended some effort in trying to identify the colorings used by competitors. But as the next chapter will demonstrate, this was no straightforward task.

The rapidly expanding numbers of commercial dyes on the market combined with the huge range of names used for different—and sometimes the same—dyes, together with technical difficulties of testing and the absence of rigorous and standardized tests, made identifying the dyes almost impossible. Many of these tests carried out at Rowntree seem to have been conducted by members of the Rowntree family themselves, rather than consultant chemists. Extracts from John Wilhelm Rowntree's experimental books include assessments of Glasgow-based Hay Bros' gums that describe Hay's black currant pastilles, for example, as being a "dingy, artificial purple" while the colors for Craven's gums are described as "poor and artificial." Other gums tested included Terry's, Buchanans, Barratts, Pascalls, Taverners, Tucketts of Bristol and some gums bought in Paris and New York.[69] However, the notebooks offer no evidence that Rowntree was able to accurately identify any of the dyes used by other confectionery manufacturers.

Occasionally, food manufacturers sought help from consultant analysts to identify the colorings used in competitors' products, and even those used in their own products, as the following letter from the Reading public analyst and medical officer Alfred Ashby to the local biscuit manufacturer Huntley and Palmer shows.

> From Borough of Wokingham Urban Sanitary Authority, Town Hall, Reading, 20th May 1891. Alfred Ashby, Medical Officer of Health.
>
> Dear Mr Palmer, I had only two of the biscuits with dark red, and only three with the yellow colour, so could not make so complete an examination of those as I should have liked. I was therefore anxious to have some of the same colouring matter as had been used in them. Most of the colouring matters which are usually used are perfectly harmless, unless they happen to contain any arsenic, but that is very rarely the case now. There are, however, two yellow coal tar colours, which I believe are sometimes sold for saffron, and which have been proved to have injurious effects. I wanted to identify the yellow colours to see if by any chance either of these had been used, but the quantity at my command was not sufficient for the purpose. (I have sent my report on the biscuits).[70]

This letter shows the difficulty facing public analysts, and indeed food manufacturers, in attempting to identify the composition of the many hundreds of different types of dyes, whether coal tar or plant, animal or mineral based, being sold as "harmless" food dyes by chemists and grocers during this period.

When chemists began to be employed as staff by Rowntree in 1897, the preparation of artificial colorings became one of their first tasks. Benjamin Seebohm Rowntree, who had completed a short course in chemistry at Manchester's Owens College, had set up a basic laboratory in the factory in 1896, according to Stanley Allen, one of the first chemists employed by the company. The following year, 1897, Rowntree hired Samuel Davies, head of the chemical department at Battersea Polytechnic, to set up a chemical department at the company.[71] A few years later, in 1900, a larger general laboratory was opened. According to Allen:

> the preparation of colours for factory use had started in the old Laboratory and new ones were constantly being added. Prior to this all had been bought in paste or liquid form. We had a long job in finding fast, non-toxic, clean colours; the first into use were Auramine and Rhodamine and the most difficult were suitable substitutes for the expensive Saffron and Carmine, but these came at last. Another difficult one was to get a cachou violet that would not stain the mouth. A boy was employed to do most of this mixing work and Randles eventually took charge of it. When flavours were added to the Laboratory's manufacture, a flat roof above us was walled and roofed in and so became the first extension; and colour and flavour work were transferred to it.[72]

The above examples of food manufacturers' use and trialling of dyes suggest that chemists, including public analysts, were well aware of the economic and aesthetic values of the new coal tar dyes for food manufacture. They also show, however, the concerns and lack of consensus that remained about their use, their elusiveness, and their lack of determinacy.

The use of chemical dyes in food is one of the first examples of novel scientific processes or technological products to have become integrated into society in unexpected ways. The aniline and azo dyes were also the first of hundreds of thousands of synthesized chemicals to become embedded in commerce and everyday commodities. It is telling that food and dye manufacturers remained silent on the question of the use of synthetic dyes in foodstuffs until concerns began to be publicized in the media. At this point, both types of manufacturer turned to chemists to help them appease growing concerns and to legitimize the use of the new dyes in food and drink.

Chemists recognized the need to reassure the public over the safety of the new chemical dyes and also saw the debate as an opportunity to establish chemists and chemistry as important contributors and arbiters of the food industry. But in order to achieve this, chemists had to devise and agree on tests in order to understand and assess the dyes better. And as the next chapter will show, this was not an easy task.

4 The struggle to devise tests to detect dyes and assess their toxicity

For analysts to claim the public authority to address concerns about the use and long-term physiological effects of chemical additives, and to inform the introduction, amendment, or implementation of food legislation, they needed a reliable system of detecting these substances in the food supply. However, synthetic dyes were a totally new and unknown commodity with no prior tradition or recognition. Existing apparatus and methods of analysis had to be adapted and reformed. Although these new chemical substances were brilliantly colored and increasingly ubiquitous, they remained almost impossible to detect for decades. Ironically, while these substances were themselves created by chemists, the detection and evaluation of their use in food was a failure in analytical chemistry.

In a precursor to the risk and regulation dilemmas confronting scientists today, nineteenth-century chemists were trying to understand and arbitrate the use of new scientifically manufactured substances that were being rapidly and widely introduced into everyday life.[1] Understanding the new chemical substances involved a rethinking of testing methods and the assemblage of a wide range of new and existing practices and regimes. Individual chemists from different social and institutional settings battled to establish rules and platforms from which to know and normalize these novel chemical entities. Chemists across Europe and America had different views of the new dyes in food and drink, according to their social, political, economic, and institutional status and circumstances. Investigating the tests used to assess and detect the dyes demonstrates and contextualizes the problems that professional food chemists faced when acting as mediators between the public, politicians, and the food industry.

At the end of the nineteenth century, the German chemist Theodor Weyl argued that there were two possible approaches to controlling the use of the new dyes in food. The first would be to ban all synthetic chemical dyes in food and drink, an option he judged impractical and likely to be resisted strongly by food producers and legislators. The other way forward, he argued, would be to designate a permitted list of colors known to be harmless, with new coloring matters added to the list only after proper testing and with all permitted coloring matters being detectable, even in small quantities.[2] This latter suggestion was similar to the strategy eventually adopted in the US. Together, hygienic chemists and public analysts looked to tests and standards to create order and ensure that science, and scientists, could assert authority and integrity in a commercial world that threatened to exploit the plethora of novel dyes for financial gain, at possible risk to consumers and the public's confidence in these new scientifically created substances.

The presence of synthetic dyes in food was from the outset a contested issue among chemical analysts involved in food monitoring and production. Many of them argued that the new synthetic dyes provided a scientific way to color and preserve food with harmless substances, increasing the variety and affordability of food available to the public. Other chemists, however, claimed the safety of the new dyes was not established and that the synthetic dyes, like other artificial colorings before them, were used to disguise the quality and content of food products, thus deceiving the consumer. Aside from the issues surrounding the democratization and transparency of the food supply, chemists also remained divided on the subject of toxicity, particularly as a result of the tiny quantities of individual dyes used in food.

Nineteenth-century scientific journals in Germany, France, the US, and Britain published extensive discussion of the new dyes and experimental information, with domestic magazines printing extracts from foreign journals. Analysts and chemists also traveled overseas to learn more. Reports of successful attempts to detect synthetic coloring originated from throughout Europe and the Americas. Chemists from countries from Argentina and America to Germany, Italy, Spain, and France were devising experiments to detect coal tar colors in food.[3] The journal articles indicate how extensively synthetic dyes were being employed in food production throughout Europe and America by the late nineteenth century, as well as the difficulties analysts had keeping track of their use. However, it is clear that there was no widespread agreement as to how much testing for the dyes was necessary. Even among those who did agree about the need for tests, there was no consensus about which tests were best. Experimenters employed an array of analytical techniques to identify these new substances, from distillation, heating, filtration, treatment with acidic reagents, and titration to spectroscopy and microscopy, relying extensively on their senses of smell, taste, and sight. They adapted and blended methodology and techniques from traditional analytical chemistry with those of existing industries and crafts, particularly textiles and dyemaking, together with new chemical and industrial techniques and emerging concepts of structural and synthetic chemistry.[4] Reaching consensus over testing methods as to how the dyes could be known and understood was critical to policing their use in the public domain, as well securing the status of chemists and chemistry. German chemists, especially those involved in the developing fields of hygiene and physiology, played a central role in testing the new dyes, assessing their suitability for consumption and ultimately legitimizing them. Physiological experiments, such as toxicological testing, and the new discipline of histology provided alternative regimes of testing with different disciplinary histories, linked to chemistry since the early nineteenth century.

Chemists were faced with having to test for, and assess, substances that were constantly changing and proliferating and whose presence was seldom disclosed. One can test for known substances but not new, unknown, and unidentified ones. This situation still persists today, as histories of twentieth-century industrial contamination and current debates about environmental pollution demonstrate.[5] Besides not knowing what new chemicals were actually being added to food, chemists had no idea how these unknown chemicals interacted with other chemicals, either natural or synthetic, that were also present in the food, consuming body, or external environment. The interaction between different chemicals or chemical processes, known as synergy, is now recognized as exceedingly difficult to assess outside of the laboratory.[6]

The nomenclature and taxonomy both of the dyes themselves and of color as a key measurement and indicator of the new substances was far from straightforward. As experimental descriptions attest, the evaluation of subtly differing hues of color was critical to "knowing" the new dyes. With no universally agreed-upon system of color measurement, however, the use of color to categorize the dyes, or to identify them through analytical techniques such as titration, was extremely problematic. As histories of taxonomy have shown, naming and classification are a complex process, dependent on consensual agreement and starting from the bases of differentiation and categorization.[7] Whereas a system of nomenclature requires a cohesive and agreed-upon set of conventions for creating and applying names to things, a taxonomic system needs an integrated set of rules for defining how classifications are constructed. Reaching consensus over both nomenclature and taxonomy involved a transformation in how chemists viewed these new substances.[8] Evan Hepler-Smith observes that the naming of the thousands of new organic substances being synthesized in the late nineteenth century was a matter of considerable debate amongst international chemists.[9] The creators, manufacturers, wholesalers, retailers, users, and consumers of these dyes all conferred different names upon them, leading to many synonyms being used for identical chemicals. Even chemists themselves were unable to agree upon a chemical nomenclature for the new substances being discovered, prompting efforts to create a unified system of nomenclature. While French chemists promoted a system based on the chemicals' properties, German chemists, guided by Adolf von Baeyer, convinced delegates at the 1892 Geneva Nomenclature Congress to adopt a system based on chemicals' structural formula and diagrams.[10]

Just as chemists found it hard to agree on nomenclature, there was little consensus on the type of tests needed to assess the dyes. The earliest tests were based on techniques of elimination derived from classical analytical chemistry, using regimes of proof such as titration. The difficulty for the analytical chemists in using elimination tests for known additives was that, by the 1880s, there were hundreds of dyes in the European and American marketplace—some known, many unknown—and food and drink manufacturers often added several coloring additives to one product in order to make the detection of discrete dyes harder. Testing for a known substance, or to eliminate a specific substance, is a much easier task than identifying the entire range of substances that might be present in any food or drink sample. Checking for the presence of several unspecified dyes would entail a whole battery of individual tests. This new testing regime proved to be a complex epistemological maneuver comprising old repertoires of testing relocated to a new context, with collections of tests assembled by an array of chemists from different specialties and chemical practices, artisanal, technological, and medicinal.

Meanwhile, methods of testing for toxicity devised by physiological chemists using animal and/or in vitro testing was just as problematic. In these tests, animals or test tubes were used as proxies for the human digestive system. Such types of models were beset by a multitude of problems, including a large number of variables, nontransferability, and individual variation and tolerances of the consuming subject. Other problems included the difficulties of assessing long-term effects and deciding which dyes among hundreds to test for, and in what form.

Classifying chemicals and color: Rosaniline by any other name . . .

The sheer number, complexity, and confusion of names and classification of synthetic dyes in the nineteenth century are immediately apparent when reading analysts' reports from this period. Following a trip to Paris in 1885, the British public analyst John Muter described tests used by French chemists to detect the use of rosaniline, safranine, aniline violets, mauvaniline, chrysotoluidine, amidonitrobenzol, roccelline, foundation red, Bordeaux red and blue, Ponceau red and blue, Biebrich scarlet, tropaeolin, chrysoidine, helianthine, eosine B, eosine JJ, safrosine, ethyleosine, and red corallin.[11] This extensive list of just some of the new synthetic substances being used to color food and drink makes clear the difficulties analysts faced. By the early twentieth century, Bernhard C. Hesse estimated that there were 921 commercial coal tar dyes made from intermediaries from nine crude coal tar waste products, while Fritz Redlich put the total at between 1,200 and 2,000. Johann Peter Murmann suggests that the discrepancy in the figures could be caused by the fact that the dyes were sold under different trade names, with hundreds of dyes continually entering and leaving the market. Murmann quotes Hesse's observation that by 1897 the existing dye molecules were sold under 8,000 different trade names.[12]

Some names were given to dyes because of their chemical construction, some described their color such as "butteryellow," while others referenced the dye's inventor or place of manufacture such as "Martius brown" and "Manchester brown." Two—magenta and solferino—even celebrated victories of Napoleon III. These names, like Lyons blue and French red, had popular appeal in the fashion industry, where high-spending consumers looked to Paris and France for the latest trends. Similarly, British dye manufacturers promoted colors such as regina purple and Britannia and imperial violet.[13] Many of the earliest aniline dyes were named using classical terms for colors (*flavin*, yellow); flowers (*safranin*, saffron; *fuchsine*, fuchsia; *mauvein*, mallow; *rhodamin*, rose); minerals (*auramin*); or animal dyes (*purpurin*). Later names, such as *Methylviolett* or *Benzingelb* tended to indicate their chemical composition, applications, or properties. As the number of dyes continued to expand, letters indicating coloration were added such as R for "*rot*" (red), G for "*gelb*" (yellow) or *grün* (green), etc. Some dyes were named in groups, sometimes linked to the manufacturer or a brand name.[14] In many ways the profusion of chemical dye names, with their different origins and associations, was comparable to the situation in botany before the Linnaean classification system was adopted.[15]

Moreover, chemically identical dyes had different names in different countries or when sold by different wholesalers. Manufacturers would often keep the chemical formulae of their new dyes secret, leading to uncertainty and confusion, while producers and retailers frequently reused established names for commercial reasons, resulting in the same name being applied to several different types of dye. So, for example, chrome yellow, Turner's yellow, yellow lead oxide, and some azo dyes, among others, were all called *Neugelb*, while *Kaisergelb* was used for yellow ochre, cadmium and chrome yellow, and Aurantia, a nitro dye.[16]

According to Weyl, "the trade names of the coal tar colors are mostly fanciful, since the scientific titles are cumbersome and difficult to remember." For example, aurantia's technical name was hexanitrodiphenylamine. Weyl complained that colors were given different names by different suppliers, noting that crocein orange, ponceaux 4 GB, and brilliant orange were all the same color, while Bismarck and Manchester brown, phenylene brown, and canella were also identical dyes. Weyl had no doubt that this practice led to confusion in the market, which traders used to their advantage:

> Different colors are often designated by the same name, especially with a view of substituting a cheap for a costly product. In this way . . . the low priced Martius' yellow is called naphthol yellow S, although the latter name belongs to a more expensive preparation. Finally, mixtures of familiar colors necessary to produce peculiar tints are frequently sold under new names, with deceptive intent. Cardinal, for example, is a mixture of chrysoidin and fuchsin.[17]

This was a deception that could harm health as well as quality and commerce, since Weyl's experiments indicated that Martius yellow, for example, was toxic, whereas napthol yellow S could be safely consumed in small doses. The German bacteriologist Ferdinand Hueppe, who, like many German scientists, increasingly used the new dyes to stain tissue samples, also highlighted this situation. Hueppe noted that different trade names were used for identical preparations. Worse still, according to Hueppe, was the secrecy surrounding the dyes and spurious statements made by their manufacturers to deliberately mislead their competitors. Hueppe studied medicine in Berlin and later worked with the bacteriologist Robert Koch in Berlin (1880–84), where Weyl had also worked, and then at the Carl Remigius Fresenius Institute in Wiesbaden. Hueppe was also involved in the German hygienic movement, writing on vegetarianism and the importance of physical exercise, and became the first president of the German Football Association (1900–1904). Hueppe was an early proponent of hormesis, the theory that a small amount of a toxic substance can in some circumstances stimulate bacteria, while a high dose of the same substance inhibits growth of the same bacteria.[18]

Classifying the new dyes

In order to reach any scientific consensus concerning tests for the presence of dyes in food and to determine the toxicity of individual dyes, chemists sought uniform systems of classifying both dyes and colors. In 1888, two German chemists, Gustav Schultz and Paul Julius, published *Tabellarische Übersicht der künstlichen organischen Farbstoffe* (A systematic survey of the organic dyestuffs). This work was later translated and adapted into English by the British chemist Arthur G. Green.[19] In the translation, which came to be known in the English-speaking world as Green's Table, nomenclature, classification, and formulae had been changed. While chemists sought increasingly to classify substances according to their chemical formulae, the understanding and theories of structural chemistry varied among different communities of chemists, resulting in differing chemical formulae and ways of describing and cataloguing identical compounds and substances.[20]

All three authors worked in the dye industry. Schultz was head of research and development for the Berlin-based Aktien-Gesellschaft für Anilin-Fabriketen (AGFA). Julius was a senior chemist with Badische Anilin und Soda Fabrik (BASF) in Ludwigshafen. Green was chief chemist to the color works of the Clayton Aniline Co. Ltd. in Manchester, professor of tinctorial chemistry at the University of Leeds, and examiner in coal tar products for the City and Guilds of London Institute. As chemists in manufacturing dye companies, all three would have been acutely aware of the difficulties of classifying dyes and determining their molecular determination. Both of the German chemists had themselves taken out patents for various dyes. The industry was awash with disagreements as to the composition, classification, and manufacture of patented dyes. Moreover, Schultz became involved in the infamous Congo red patent trial between Bayer and AGFA in 1889.[21]

The author of any classificatory compendium was also faced with the continual appearance of new dyes and abandonment of old ones.[22] Of the 454 coloring matters in Green's first edition of 1894, 59 had become obsolete by the second edition in 1904, and 300 new ones had been added by 1908, at which point the tables included 695 artificial coloring matters. Green pointed out how proliferation was additionally problematized by the "increasing number of dyestuffs, mostly of very recent introduction, concerning the preparation and constitution of which but little is known, whilst the manufacturers evince a very natural disinclination to furnish particulars regarding them."[23] He classified the synthetic dyes into twenty different groups, including (1) nitro coloring matters; (2) monazo; (3) diazo; (4) triazo through to (9) diphenylmethane, and (10) triphenylmethane; and (19) quinolone and (20) sulphide.[24] The table of coloring materials listed each dye's commercial and scientific names, chemical formula, method of preparation, year of discovery and discoverer, manufacturer, patent details, and applications. Each dye was assigned a specific number, subsequently to become known as its Green number. Comments in the book about the chemical formulae used to describe the different dyes suggest that, even by 1908, chemists had not reached any consensus for the use of standardized formulae for dyes.

Although many German and British compendia of dyes were published in the nineteenth century, the Schultz and Green guides became the most widely used publications. Dye historian Brian Burdett argues that the comprehensive nature of both publications and their tabular form and use of classification numbers helped promote their use.[25] The Schultz guide proved popular with dyers and colorists, while chemical analysts, especially in Britain and North America, tended to use the Green guide, which included more details on chemical reactions, permitting the identification of individual dyes. However, the debate over classification and whether it should be based primarily on chemical structure, color, application, manufacturer, manufacturing process, date of invention, colorfastness, or any other criteria continued throughout the twentieth century, as many more new dyes entered the marketplace. In 1922, the British Society of Dyers and Colourists published a new color index based on the chemical composition of each dye. This eventually became a standard classification reference after it was amalgamated in the 1950s with the classification used by the American Association of Textile Chemists and Colorists. Both organizations based their classifications on the Schultz tables.[26]

The difficulties of measuring color

Reaching agreement on how to measure gradations and different hues of color was an extremely problematic issue. Color wheels, with graduated hues and degrees of color, had been created and used by artists and natural philosophers, including Newton and Goethe, since the seventeenth century. However, the naming and categorization of separate colors in a universally agreed-upon system proved an intractable problem. Even today, many different color measurement systems are used for different applications.[27]

In 1905 the American artist Albert H. Munsell compiled and published a widely adopted color classification system, based on the color sphere created by the German artist Philipp Otto Runge in 1809 and the 1839 hemisphere of French industrial chemist and colorist Michel-Eugène Chevreul, director of dyeing at the Gobelins tapestry manufactory. Munsell's color sphere was an extension of earlier color wheels, which measured hue and brightness, adding color saturation, or chroma, as a third dimension. Munsell based his measurements on hundreds of individual human visual responses to different hues, brightnesses, and saturation levels. A professor of art, Munsell designed his system as a teaching aid for his students, but it eventually was adopted as a classification system in many different types of color measurements including hair and skin colors in forensic pathology, soil colors, and dental restorations and for assessing the color and quality of beer.[28]

Separately, printers and dyemakers, including the US printer Louis Prang and the games manufacturer Milton Bradley, produced individual color charts and classification systems for their own use and that of clients.[29] Other industrialists also began to develop separate empirical measurement systems for their own types of products, based on color comparisons with standard samples. In the 1880s, a British brewer, Joseph Lovibond, devised a "Tintometer" to assess the color of beer against a graded set of colored glass filters.[30] This system was widely adopted, particularly in industries like food and drink that produced fluids such as beer and oil. However, it was not without its problems, particularly relating to replication and standardization, leading several American industry users, including the American Oil Chemists Society, to ask the US National Bureau of Standards to standardize the glass used in Tintometers in the 1920s.[31]

In the nineteenth century, the science of color became a new field of research bringing together artists, colorists, physicists, chemists, physiologists, and practitioners of new sciences such as pycology and anthropology.[32] Colorimetry, the measurement of color, proved an intractable issue for much of the twentieth century. Many competing systems of measurement remain on the market today, including both the Munsell Color System and Lovibond spectrophotometric colorimeters. The US National Bureau of Standards began to tackle the problem of standardizing the measurement of color soon after its formation in 1901, and frequently found itself acting as arbiter in disputes, such as in 1912, when the butter, oleomargarine, and cottonseed oil industries asked for help in color grading their products.[33] The historian of science Sean Johnston argues that reaching agreement over the measurement of color proved far more complex than over the measurement of light.[34] Indeed, as nineteenth-century artists and scientists showed, the perception of color is subject to many variables, including what other colors are situated nearby, what type of light the color is exposed to, and the type of material that is colored. During the first few decades of the twentieth century, committees of physicists, psychologists, industrialists, and artists debated the best way to measure color, with each user group eventually choosing its own preferred classification systems.[35] Sociologists of science have observed that consensus is often harder to obtain among a wider field of interests, cultures, and disciplines. Raymond Cochrane notes in his history of the Bureau of Standards that "the field of research in the Bureau in which undoubtedly the greatest variety of industries and interests had an initial concern was the standardization of color."[36] It is no surprise then that the calibration and measurement of color was a problem for analytical chemists.

The ontology, epistemology, and social and technological circumstances of measurement and the concept of measurement continue to attract interest among philosophers, historians, and sociologists.[37] Empirical and experimental constraints all inform the choices of measurement standards and techniques of testing, while the forms of measurement that are adopted depend on consensus.[38] Hasok Chang argues that constructing a concept of quantity, and measuring it, is an iterative and codependent undertaking, involving a continual adjustment of conventions.

Chemical analysis

The ability to differentiate between, and assess, colors was a critical part of the analytical techniques used by chemists to identify whether, and which, synthetic chemical dyes were being used in food and drink. Analytical chemistry had become an established part of chemistry during the eighteenth century with a wide range of quantitative and qualitative techniques developing throughout the first half of the nineteenth century. The development of new analytical techniques and equipment for wet, or classical, chemistry during the eighteenth century led to a transformation of the chemist's laboratory by the nineteenth. Aspiring chemists could purchase an analytical laboratory that could be set up at home, in the field, or in a shop, a development that led to an expansion of practical chemistry and education and the beginnings of a chemical profession. Chemical analysts, who had already been testing mineral waters and metals in the eighteenth century, were increasingly called upon to test everything from drinking water to food, air, rivers, and dead bodies. But while analytical chemistry helped chemists find a recognized and reputable role in society, their techniques, experiments, and diagnoses remained highly contested even among their peers.[39]

The early tests adopted by chemical analysts for identifying synthetic dyes derived from classical analytical and titration methods, using chemical reactions and color changes to determine different substances in solutions, techniques well-established by the mid-nineteenth century. Titration or volumetric analysis developed in the eighteenth century as a quick and efficient quality test for chemicals such as sulphuric acid, hydrochloric acid, and soda being produced for use in the textile and other industries. If the presence, quantity, or purity of a substance needed to be established, another substance with which it reacted was added, known as the reagent. The reaction needed to produce a marker or form of identification, such as the formation of bubbles or a color change. Since at least the sixteenth century, dyes had been used in such tests as an indicator, as they would change color according to the acidity or other properties of the substances being tested.[40] However, while titration methods were frequently used in the analysis of industrial chemicals, the use of similar techniques for detecting dyes in food and drink was limited. The types of test adopted in cases of suspected food adulteration often stemmed from the analyst's own background or knowledge base as well as his interpretation of the dyes themselves.

Much of the earliest experimentation devised to detect the presence of synthetic dyes in food and drink took place in France, a country with a well-established tradition of analytical chemistry and where the bright new synthetic colors, particularly fuchsine, proved a cheap method of enriching the color of wine.[41] In 1876, the French chemist Armand Gautier observed that the use of artificial colors in wine was "an increasing practice" and that while vegetable dyes such as beetroot and elderberries were employed, aniline dyes, especially fuchsine and grenat, a by-product of fuchsine, were increasingly being used.[42] Gautier was abreast of developments in structural and synthetic chemistry, including the new chemical dyes, and had been a student of Adolphe Wurtz, the pioneering French structural chemist and professor of chemistry at the Faculté de Médecine in Paris and later of the Sorbonne.[43]

Gautier was concerned about adulteration with fuchsines, aniline reds, and violets "used in large quantities either alone or mixed with various other yellow or red substances to diminish the brilliancy of their tint" as well as about "substances sold under fantastic names, such as 'colouring,' 'caramel,' or 'colouring fluid,'" consisting of fuchsine residues with beetroot, used to mask the presence of other adulterants. He noted that the prevailing methods to detect the presence of artificial colors in wine relied on the "fugitive" nature of the dyes. After providing a review of the various methods devised by other chemists for using different reagents, such as sulphuric acid, barium peroxide, ammonia, or borax, to detect chemical dyes in wine, he condemned them all as "valueless." Instead, he devised a systematic sequence of elimination tests to be used on wine after purification with diluted albumen (egg white), filtration, and sodium bicarbonate.

Table 4.1 summarizes Gautier's test sequence for detecting fuchsine. This short example demonstrates the multiple maneuvers that chemists had to undertake in order to detect and identify the presence of just one coloring substance in wine. It also shows how negotiable the science of detecting organic substances was, and the large part played by tacit skills and experience: much of the diagnosis revolved around the identification of only subtly differing hues of color. In the absence of a standardized color nomenclature, this was a complex and uncertain process, as in the description of a liquid becoming lilac or violet or sometimes "only winey, or dashed with violet." Many of the tests were older and had served to detect natural dyes long used in wine such as whortleberries, logwood, or black elder. The methods proposed by Gautier for detecting the new aniline dyes were an extension of traditional analytical methods coupled with the isolation methods used by the exponents of the new synthetic experiments such as Hofmann. Elimination tests have a long history in chemistry, and Gautier's maneuvers incorporated traditional chemical techniques from industry and artisanal practices including distillation and titration. Gautier pointed out the value of his proposed systematic approach, on the basis that adulterated wine was likely to contain more than one type of adulterant, recommending that

Table 4.1. Gautier's elimination process to detect fuchsine in wine (omitting the stages not involving the detection of fuchsine)

A	If, after washing, the precipitate remains wine colored, lilac, or maroon, this shows it is either natural wine or adulterated with the usual adulterants, excepting indigo, so proceed to C. If the precipitate is very deep wine color, violet blue, or bluish, it shows the wine to be from the deepest colored grapes or dyed with indigo. Proceed to B.
B.	Wash the precipitate with water, then with alcohol of 25 percent, part of which is removed and boiled with alcohol of 85 percent. If the filtrate is rose or wine colored, saturate a portion with dilute potassium carbonate. If the color changes to brown or blackish brown, it demonstrates that the wine is either natural or adulterated with substances other than indigo. Proceed to C. If the filtrate is blue, treat part of it with dilute potassium carbonate. If the deep blue liquid then changes to yellow it indicates the use of indigo.
C.	Treat two cc of wine with 6–8 cc of a 1/200 solution of sodium carbonate. If the liquid becomes lilac or violet ("sometimes the liquid becomes only winey, or dashed with violet"), it indicates the use of brazil wood, cochineal, Portugal berries, fuchsine . . . wines of certain sorts, fresh beetroot, logwood, both elders, whortleberries, . . . Proceed to D. If the liquid becomes bluish-green, sometimes with a faint lilac tint, it indicates natural wine or the presence of hollyhock, privet whortleberries, logwood, Portugal berries, or fuchsine. Proceed to M. If the liquid becomes greenish-yellow, without any blue or violet, it indicates beetroot, whortleberries, or certain rare varieties of wine. Proceed to L.
D.	The liquid from C is heated to boiling. If the liquid remains wine-violet, rose, or wine-lilac, or becomes a brighter lilac, it indicates the presence of logwood, Brazil wood, cochineal, or certain varieties of wine. Proceed to E. If the color disappears or changes to a yellow, maroon, or reddish tint, this indicates wine, fuchsine, elders, whortleberries, Portugal berries, or fresh beetroot. Proceed to F.
F.	Treat 4 cc of the wine with alum and sodium carbonate and add two or three drops of very dilute sodium carbonate and filter. If the filtrate is lilac or winey, it indicates Portugal berries or fresh beetroot. If it is bottle-green or reddish-green, it indicates natural wine, fuchsine, black-elder, whortleberries, or beetroot. Proceed to H.
H.	If the *alum-lake* obtained from F (bottle-green) is bluish-green, green, or faintly rose-tinted, it indicates natural wine, whortleberries, beetroot, or fuchsine. Proceed to J.
J.	Treat 5 cc of the clarified wine with a slight excess of ammonia, heat to boiling, and after cooling shake with 10 cc of ether, decant and evaporate the ether, and treat the residue left on evaporation with acetic acid. If the liquid becomes red, fuchsine is present.

Source: A. Gautier, "On the Fraudulent Colouration of Wines," *Analyst* 7, 1876, 130–31.

fuchsine "should be sought in all wines found to be adulterated with other substances."[44]

In the task of identifying the presence of any one of hundreds of new substances used in minute traces for which there were no known reagents or positive tests, chemists resorted to tests derived from domains where the effects of certain dyes were already established, such as the textile and food industries. So, for example, in a simpler stand-alone test for detecting fuchsine, Gautier employed a different technique entirely borrowed from practices of the textile industry:

Fuchsine rapidly separates from the wine to which it has been added. A skein of silk becomes dyed rose by soaking in a wine adulterated with fuchsine and its colour passes to yellow on treatment with hydrochloric acid, but to bright red, if the wine was pure. The dyed skein treated with dilute cupric acetate, and dried at 100 ° becomes fine deep rose-violet if fuchsine is present, and of a lilac tinged with ash-grey if the wine is pure. This re-action is very sensitive.

Gautier also described a "tolerably successful method" that involved putting skeins of silk or wool treated with mordants such as alum with cream of tartar, oxychloride of tin, and acetate of alumina into the suspected wine. After fixing the color from the wine onto the skeins, and then applying different reagents such as "ammonia, lime-water, chlorides of zinc, iron, calcium, salts of copper and tin. . . . new re-actions were observed, which are characteristic of certain colouring matters."[45] Gautier was establishing and promoting his own techniques by coupling classical analytical methods of titration with practices from the textile trade, where the use of dyes was familiar and understood.

Similar assemblages of different strands of knowledge and practices can be gleaned from a description of other tests devised by French chemists. The British analyst John Muter published details of these tests after his visit to the municipal laboratory of Paris in 1885, one of the main food adulteration test centers in France.[46] Muter listed three different tests used to detect aniline dyes in wine:

1. Sticks of chalk "like those used for blackboard writing" were steeped in a 10 percent aqueous solution of egg white and treated. Two drops of wine were poured on the porous surface, which went either grey in the case of ordinary wine or bluish in the case of a young wine, but without any trace of green, violet, or rose, unless it was artificially colored.

This test, like Gautier's, relied on the fact that aniline dyes fasten to animal protein, hence the traditional use of egg white, a procedure that would have been appreciated in the textile industry, where aniline-based dyes were known to be effective on animal-based textiles such as wool and silk.

2. Some wine was mixed with baryta water until it became greenish, then it was shaken with acetic ether or amylic

alcohol and allowed to settle, taking care that the mixture
be slightly alkaline. Pure wine gave no color in the upper
layer, while coal tar derivatives of a "basic nature" gave
various colors, indicating that research should be carried
out for such substances as amidobenzene, fuchsine, safranine, chrysoidine, chrysaniline, mauveine, methyl-violet,
and beibrich red.

This experiment was a good example of how the practice of color
changes obtained in centuries-old titration tests was combined with the
practice of nineteenth-century synthesis of new substances.

3. Ten cc of wine was rendered strongly green by the addition
of 2 or more cc of 5 percent potassium hydrate. Add the
same amount of a 20 percent solution of mercuric acetate,
and filter after shaking. With pure wine the filtrate itself
was colorless after treating with hydrochloric acid, while
that containing any coal tar derivative of an acid nature
was colored red or yellow.

This test indicated that while the analyst could detect the presence of a
"basic" acid dye, the identification of specific chemical substances was
a very different and more difficult task.

Muter made no comments on the efficacy of the French experiments,
but suggested that wine industry operatives remained one step ahead of
France's municipal analysts, the French version of Britain's public analysts. According to him, the new synthetic dyes had rendered the "use
of the old-fashioned logwood and other vegetable colours . . . a thing
of the past" with two new mixtures being sold for deceiving French
analysts, one being a "mixture of amido-benzene, methyl-violet and the
acid sulpho derivative of fuchsine. . . . (and) a compound of methylene blue, diphenylamine range, and the acid sulpho derivative of fuchsine. . . . sold in commerce as Bordeaux Verdissant."[47]

Historians have shown the difficulties faced by French chemists in
trying to identify the dyestuffs being used in the wine industry, despite
the fact that, by the 1870s, France enjoyed extensive links between its
regional universities, agricultural research stations, and local industries
and agricultural producers. The ability of wine producers and merchants to elude analysts' attempts to detect the false coloration of wine
continued throughout the nineteenth century, despite the establishment
in 1878 of the Municipal Chemical Laboratory in Paris, set up primar-

ily to detect artificial coloration of wine and other food adulterations. This new laboratory was equipped with darkrooms for polarimetry and spectroscopy, as well as more traditional analytical apparatus. The laboratory maintained close links with the Toxicology Laboratory of Paris and was headed until 1911 by Charles Girard, a founder of the Saint-Germain aniline dye manufacturer La Fuchsine. The fact that the French municipal chemists, with their advanced equipment and good connections, were still unable to detect and specify the presence of individual aniline dyes in wine reliably highlights the difficulties that all public analysts faced.[48]

The constant battle between the food and drink manufacturers who used the synthetic dyes and the chemists tasked with detecting them was depicted by the German chemist Joseph Herz when recounting a series of experiments for detecting artificial colorings conducted at the Würzburg Research Institute for Food and Drink (Untersuchungsanstalt für Nahrungs- und Genußmittel zu Würzburg).[49] Herz noted that because magenta was the coloring matter most looked for by food analysts, it became the coloring least used. Herz's experiments used filtration and a series of reagents to eliminate and detect a range of substances, including rosaniline sulphonate, orseille, ponceau colors, cassinine, vinicoline Bordelaise, and Bordeaux B (in powder form), by observing the different colors obtained. As with other experiments, much reliance was placed on the visual and subjective observation of differing shades of color. As the discussion above on the impossibility of generating a standardized color measurement framework suggests, however, grading color for identification purposes relied heavily on tacit and empirical skills of the chemist, an unmanageable way to differentiate between 695 dyes![50]

Another test depending upon empirical and tacit knowledge was published in the German weekly chemical publication *Repertorium der analytischen Chemie für Handel, Gewerbe und öffentlich Gesundheitsplege.* In an 1886 report, K. Fleck observed that "many makers of farinaceous foods are in the habit of colouring them" with binitrocresol, or Victoria yellow. However, because manufacturers of products such as macaroni were also in the habit of using picric acid (also known as Welter's bitter, with the chemical name trinitrophenol), the author described a method of detection for the poisonous acid that involved mixing the powdered food with hydrochloric acid and introducing zinc. Throughout the process the tester was advised to test for the bitter taste of picric acid.[51]

Taste, color, and smell remained an important part of the testing process even where experiments in synthesizing new chemicals, pioneered by Gautier, Wurtz, and Hofmann, were adapted to look for chemical coloring in food and drink. Charles O. Curtman used the "well-known isonitril test" devised by fellow Liebig student Hofmann in 1867, to detect primary amines to test for aniline colors in wines and fruit juices. According to Curtman, this test was "successful with aniline blue, aniline purple, aniline violet, magenta and ponceau red, and many yellow and green anilines."[52] Curtman, a native of Giessen, had studied with Liebig, Hofmann's mentor, before moving to Antwerp, where he worked as an industrial chemist, producing acetic acid. He subsequently emigrated to the US, later becoming professor of chemistry at the Missouri Medical College.[53]

Hofmann had synthesized isonitrils not long after Gautier had isolated them while conducting experimental work under Wurtz.[54] In his test, Curtman simply reversed Hofmann's synthesis, mixing 4 cc of wine, colored with fuchsine, with 4 cc of potash, and 2 drops of chloroform. After "gently warming for a minute and subsequent boiling, the characteristic smell of isonitril" should be perceptible, according to Curtman, who noted that the test could be made more sensitive by adding sulphuric acid.[55] This test demonstrates the value of reversing processes and reactions and shows that chemists during this period did not distinguish between synthesis and analysis in devising tests to detect the new synthesized chemicals.

All these tests and approaches demonstrate how chemists were adopting and adapting various forms of practice from different expert traditions, trying to extend existing expertise into new territory in order to detect unknown substances. This situation was similar to the problems chemists had experienced earlier in the century in trying to pin down the identity of alkaloids, a strategy that similarly depended on pulling together different forms of expertise from separate specialties.[56]

Physical analysis: jellies and microscopes

While French and German chemists were busy blending techniques from classical analysis and the new synthetical experiments to find ways of detecting the aniline dyes, British analysts took a less analytical and more down-to-earth approach to the issue of chemical dyes in food. One of the most commonly cited British techniques relied on a physical examination of staining and the use of microscopes. One of the favored methods used for detecting artificial coloration relied on the physical properties and characteristics displayed by different dyes in jelly. Gelatin is a form of edible protein extracted from animal bones and tissue, and aniline dyes adhere rapidly and strongly to animal protein, hence their use in dyeing silks and wools and as a dye source in histology. During this period, brightly colored and highly ornate jellies were becoming increasingly popular in Victorian dining rooms, encouraged by industrially produced gelatin and jelly molds.

In an address before the Society of Public Analysts at Burlington House in 1877, its president, August Dupré, dismissed "sensational statements" in the media about the "fraudulent colouration of wine," pointing out that he had never come across a bottle of red wine containing aniline dyes.[57] It is worth noting, however, that without an agreed-upon test for fuchsine and other aniline dyes, such colorings were unlikely to be detected! While Dupré played down the issue of wine being falsely colored, he did concede that "methods for the detection of foreign colouring matters in wine have to be devised, were it only to allay the fears of the public." Dupré suggested putting a cube of jelly into the suspected wine for twenty-four to forty-eight hours and observing the color and penetration of the color in the jelly, both visually and spectroscopically, to determine whether the coloring was artificial or not.[58] Rosaniline, for example, "imparts to the jelly a beautiful red colour," he observed. Meanwhile, the action of dilute ammonia on the jelly "will yield characteristic results, such as decolourising the rosaniline," he added. While uncolored wine imparted its color into the jelly very slowly, the color of wine dyed with additives such as rosaniline penetrated the jelly quicker and more deeply. Dupré urged fellow analysts to take up the challenge and devise their own experiments for detecting the false coloration of wine.[59] His comments indicate that by the mid-1870s there was as yet no standardization of tests and that the field was wide open for analysts. Later reports published in the *Analyst* suggest that British analysts held Dupré's jelly technique in high regard for some time, with Muter noting nearly a decade later that the French "seem to have missed Dr Dupré's excellent preliminary test with gelatine cubes" for detecting artificial coloring matters.[60] As in the experiments being devised by French analysts, Dupré's test coupled the use of reagents, a technique used by chemists in analytical and synthetical experiments, with an understanding of the synthetic dyes' ability to adhere to protein, a fact exploited industrially in the textile dye trade and by a growing number of scientists, including physiological and hygienic chemists. The capacity of dyes, particularly aniline and azo dyes, to stain protein-based animal tissue and cells preferentially was particularly exploited by histologists working in Germany, including Robert Koch, Carl Weigert, Joseph Gerlich, and Paul Ehrlich, who were starting to use synthetic staining in their investigation of tissue matter under the microscope.[61] Indeed, it was physiological and hygienic chemists who devised many of the tests for investigating the toxicity of the dyes.

Although the Dutch microscopist Anthonie van Leeuwenhoek had used saffron to stain histological specimens as early as the seventeenth century, the production of synthetic dyes in the late nineteenth century led to a rapid expansion in the use of dyes for staining tissue matter, including food samples, being investigated under the microscope.[62] The use of dyes as stains for making the invisible "visible" under a microscope had thus had a long history, while the microscope itself had also long been an important tool in the armory of public analysts. Arthur Hill Hassall, for example, was an early exponent of the benefits of using the microscope to identify food adulteration. The German chemist Otto Schweissinger combined the two developing traditions by advocating the use of both analytical techniques and microscopic detection of coloring matters in sausages. According to Schweissinger, the presence of magenta "is detected by cutting up the sausage into small pieces, extracting with alcohol, evaporating the extract down, and fixing the colouring matter on wool, either directly, or after taking it up with water." If this test, deriving from textile dyeing, was negative, but the use of synthetic coloring matter was still suspected, Schweissinger suggested that "the microscopic investigation of suspected particles should be resorted to." Schweissinger's approach to food adulteration appealed both to the traditions of the indigenous textile trade and to contemporaneous developments in German histology and synthetic experiments, as well as invoking the importance of sausages in local German cuisine.[63]

Testing for toxicity

While establishing consensus over experimental procedure, classification, and standards for the detection of synthetic dyes proved difficult, reaching agreement about how to determine the physiological impact of synthetic dyes on consumers proved even more problematic. This is still an unresolved issue in the twenty-first century, when the safety of dyes remains controversial and is addressed differently according to location.[64]

During the nineteenth century, a new program of experimental physiology developed, with increasing research into the effects of diet as well as poisons on the human body, together with a greater understanding of the digestive process. Much of this research centered on animal experimentation. A detailed review of scientific journals and books published during the late nineteenth century suggests that French, German, and American scientists were more active in testing the effect of synthetic dyes on animals than their British counterparts, who faced more marked public and political opposition to vivisection.[65] However, establishing experimental agreement and consensus in the new field of nutritional physiology proved problematic. Differing local cultures, laboratory equipment, and experimental skills and techniques impeded attempts to standardize tests and interpret them.[66]

British chemists even resorted to consuming the dyes themselves to test them. When Walter William Fisher, the public analyst for Oxford, Berkshire, and Buckinghamshire and a chemical demonstrator at Oxford University, identified rhodamine in sweets, he persuaded his assistants to test its toxicity by consuming "two or three grains" of the dye, enough to "colour an enormous amount of sweets." Fortunately, they suffered "no obvious effect."[67] John Attfield, editor of the *British Pharmacopoeia* and professor of practical chemistry to the Pharmaceutical Society of Great Britain, also learned how to use stomach tube technology, regularly testing the toxicity of dyes by consuming them himself. This practice convinced him of the harmlessness of most dyes:

> The question that has commonly been put to me by wholesale confectioners has been—is this colouring matter of which we show you a sample injurious to health? Up to ten years ago I had always answered that question in this way—I am not qualified to form an opinion; but since I have used the stomach tube I have endeavoured to answer the question by the simple method of taking quantities of the colouring matter, and I have never come across a colouring matter that was injurious, that was distinctly harmful, and that distinctly produced any discomfort in any way.[68]

Attfield's testimony revealed that some manufacturers had been approaching chemists for advice on the dyes they were using for at least a decade and that the analysts were not really pronouncing on their safety with any certainty. Attfield's taste-and-tell approach and his confident view that the dyes were not harmful contrasts with the detailed yet inconclusive European and American digestion experiments outlined below.

In continental Europe and the US, chemists, physiologists, pharmacologists, and physicians interested in the medical applications of the new organic compounds and substances led the way in investigating the impact of benzene and phenols and their derivatives such as aniline and other compounds on the human physiology. These scientists effectively used animal and human organisms as chemical reagents, studying physiological responses, such as the effects of different organic substances on urine, as well as exploiting the ability of the new dyes to make visible cell and color changes in the body at a microscopic level.[69] However, despite the growing number of chemists studying the physiological effects of aniline, consensus over the dye's toxicity was difficult to achieve. While Friedrich Wöhler claimed that aniline had no poisonous effect on dogs, other chemists contended that pure aniline was a powerful poison, affecting the blood, nervous, and digestive systems.[70]

Perhaps not surprisingly, given the early concern in France surrounding the use of synthetic colors in wine as well as the pioneering work of Magendie and Bernard, some of the first physiologists to look specifically at the effects of the new dyes being introduced into food and drink were in France.[71] Experiments by Georges Bergeron and J. Cloüet of the Société Industrielle de Rouen showed that dogs were able to bear doses of 20 grams of fuchsine without injury, while a man consumed 3.5 grams in a week with no effects, prompting the chemists to conclude that it was not harmful.[72] Subsequent experiments by the Rouen medical professors V. Feltz and E. Ritter showed, however, that aniline dyes had deleterious effects on both humans and dogs. Other chemists who claimed that the harm was caused by arsenical contamination of the dyes dismissed these results.[73] French chemists such as Paul Cazeneuve believed that the main danger with the new dyes lay with their arsenical contamination and the use of toxic material as mordants. In the mid-1880s, Cazeneuve and colleagues at the Lyon medical faculty performed extensive experiments on dogs to investigate the physiological effect of synthetic dyes. Most of their experiments showed that the new artificial dyes were not harmful when consumed in small quantities. Food dyes, they argued, should be pure or combined or treated with substances known to be nonpoisonous and should not be used in wine, vinegar, beer, or butter, substances widely known to be artificially colored.[74]

The most influential tester of the toxicity of synthetic dyes was the German hygienic chemist Theodor Weyl. Initially, he was skeptical about the harmfulness of aniline dyes, and many of his published comments suggest that he took on the task as a way to legitimize the dyes, as well as German chemistry, physiological science, and the role of chemists in society. As an indication of the harmlessness of fuchsine and other aniline dyes, Weyl cited the work of one of the first doctors employed by a dye factory, Wilhelm Grandhomme, who had studied workers at the Messrs Lucius & Brüning dye factories of Hoechst and found that, of the fifty-two employees working with fuchsine for between three and eighteen years, none "suffered from diarrhoea, colic, or disturbance of the urinary section, although daily breathing the fuchsine dust."[75] Grandhomme also judged aniline blue, violet, and malachite green to be nonpoisonous. Weyl viewed the wider work of Grandhomme, together with the lack of injuries and disease among workers in dye factories, as evidence of the nontoxicity of pure aniline dyes.

However, a thorough analysis of Grandhomme's work and health incidents in German factories suggests, in hindsight, that the situation was not as simple or irrefutable as Weyl concluded. During the 1890s, the German doctor Ludwig Rehn noticed and reported greater than normal incidences of bladder cancer among workers in the aniline dye industry, while in 1903 more than fifty workers at the BASF plant fell seriously ill. In 1921, the Geneva-based International Labour Office published a monogram, "Cancer of the Bladder among Workers in Aniline Factories," leading to a recognized linkage between aniline production and cancer.[76] The link between occupational diseases and the use of chemicals became more apparent in the early years of the twentieth century with the rise of medical statistics following the efforts of Karl Pearson and others in epidemiology.[77] More recent work, however, has suggested that the relationship between aniline and cancer is more complex still and that other chemicals may be involved.[78] Thus, even the conclusion of a statistical link between aniline dyes and ill health can be questioned as an artifact of broader views of them as adulterants from the late nineteenth century onwards.

As a result of his own experiments and those of other chemists and doctors, Weyl judged pure fuchsine to be nontoxic as a food dye. "A critical review of the existing literature shows that trustworthy evidence of poisoning by pure aniline colors is not at hand. In all probability, the cases in which the color has been suspected, have arisen from admixture with arsenical compounds, or the use of arsenical mordants," Weyl concluded.[79] Referring to reports of injuries caused by wearing stockings dyed with corallin, Weyl also judged pure corallin to be nontoxic and, pointing out that workers in the Würtz factory in Liebenau were "in good health," concluded that "in all probability, the [skin] eruption that attended the wearing of the stockings was due to arsenical mordants." He pointed out that "German law relating to the use of injurious colors, etc., restricts the use of arsenical mordants to such proportions as will give not more than 0.002 g to 100 sq cm of the finished goods," also noting, however, that corallin was one of the dyes prohibited by the German food law of July 5, 1887, due to the fact that commercial corallin contained phenol.[80] Like many chemists of the period, Weyl's concern centered on the contamination of dyes during manufacture with other substances known to be toxic, in particular arsenic. It is interesting to observe that, despite the continuing uncertainty and lack of consensus surrounding the new dyes, the 1887 German food law and much of the experimental work by Weyl and others made use of detailed numbers, numerical ratios, relationships, and statistics. Sociologists and historians such as Theodore Porter, Alain Desrosières, and Anne Hardy have pointed out the increased use of numbers and statistics in the late nineteenth century to assess and formulate public risk and to reshape policies designed to reassure the public.[81] The increasing use of numbers suggested a level of accuracy and certainty that was never secured. Porter has claimed that scientists, like politicians, often reach for numbers to cope with problems of organization and communication, using them to imply objectivity, rigor, and universality.[82] The issues confronting the chemists were particularly vexed because very few of the substances were pure, and some dyes needed mordants to fix them. As a result, chemists were dealing with admixtures of chemicals, including known and unknown contaminants. Moreover, while the individual elements of the mixture might not be toxic, reactions between them, or with chemicals in the body, could produce harmful effects. Establishing and assessing long-term effects of repeated small doses of a toxic substance is not straightforward. While the increasing sophistication of epidemiology during this period enabled scientists to establish links between certain diseases and their causes, its use for assessing the risk of any number of small doses of chemicals in food or the environment over a long period of time is problematic, other than in occupational settings where variables and the presence of specific contaminants are more controlled.

Despite his contention that most aniline dyes probably were safe, Weyl recognized the need and benefits of testing the new chemical dyes in order to establish a consensus around their safe use. Such a program would legitimize the status of synthetic dyes, hygienic chemists, and Germany's industrial and scientific supremacy. Weyl claimed that a "review of existing literature shows that the certainty as to the unwholesomeness or toxic nature of the coal-tar colors has not yet been attained, and cannot be attained because researches with definite purposes are almost entirely wanting."[83] Weyl's interest in the impact of the new dyes on the human body was not surprising, considering his role as a hygienic and physiological chemist.[84]

One of the problems Weyl and other chemists faced was which of the hundreds of dyes on the market they should devise tests for. This issue assumed heightened importance as it became increasingly apparent that dyes considered to be of similar chemical structure produced very different effects on the animals on which they were tested. Weyl recognized that it was impossible to test for all of the dyes. But which to choose? Colors, he noted, were subject to fashion and the commercial pressures of the market. Moreover, suppliers were often reluctant to reveal their manufacturing sources, the specifications of their dyes (if indeed they knew them), or even which were their best sellers. Eventually, Weyl decided to test a few of the most-used dyes from each type of chemical group, such as nitroso colors (based on the action of nitrous acid on phenol derivatives), nitro colors (based on the action of nitric acid on benzene derivatives), and azo dyes. He also had the problem of whether to test the pure versions of the dyes or the commercial versions, which often included many known additives and unknown contaminants. Even this choice was not simple, as in some cases the contaminants introduced during manufacture, like arsenic, increased the potential toxicity of the dye, while in others the commercial dye included the addition of substances which effectively reduced toxicity. For example, pure dinitrocresol, a saffron substitute also known as Victoria yellow, Victoria orange, and aniline orange, was judged to be poisonous by Weyl but, he pointed out, most commercial forms of the dye included 30 to 40 percent of ammonium chloride, which reduced its toxicity.[85] Even inside the laboratory these new chemical substances remained elusive, with chemists having to decide which type of dye to target and what kind of test to apply. Choosing a handful of dyes to represent the hundreds of dyes on the market was an almost impossible task.

Other parameter choices faced by Weyl and other experimenters included which animals to test the dyes on; how large a dose to give, how regularly, for how long a period and how to administer it (through the mouth or esophagus, hypodermically by injection or endermically on the skin); as well as which physical reactions to tests to diagnose poisoning (body temperature, pulse rate, respiration, etc.). Turning the animal body into something useful, particularly a proxy for the human body during the nineteenth century, has been the subject of recent discussion among historians of science.[86] The variables, starting with those among the commercial dyes themselves, were manifold, which may partly explain the lack of consensus among scientists as to which dyes were to be considered harmful.

Most of Weyl's tests were on animals of different but specified weights, usually dogs, with variable quantified doses being administered in several different ways. For example, Weyl tested dinitrocresol on both rabbits and dogs, using samples provided by both Carl A. Martius of Berlin and Schuster & Co. of Leipzig. Rabbits were administered 0.25 gram of dinitrocresol salt, dissolved in a small quantity of water, by esophageal tube. The rabbits began to respire rapidly, then quieted down before falling and dragging their hind legs, after which their pupils dilated and their extremities twitched, eventually resulting in spasms and death. Postmortems showed yellow staining in the stomach and congested lungs and liver. Doses given to dogs, of various weights, also by esophageal tube, produced vomiting and diarrhea, which, having expelled the poison, resulted in a return of the dog's appetite. Residual poison, however, eventually led to spasms and death. Weyl concluded that a toxic dose for dogs was 7–10 mg per kg of body weight for doses administered by injection hypodermically, while the fatal dose was 16 mg for completely pure potassium dinitrocresol, and 29 mg for a commercial preparation containing about 30 percent of ammonium chloride. Fatal doses administered by mouth could not be accurately monitored because of vomiting. According to Weyl, "the experiments given indicate the poisonous nature of dinitrocresol. They justify the hope that the state may restrict the sale of so dangerous a material and prohibit its employment for the colouring of food and drink."[87]

Similar experimental results led Weyl to conclude that Martius yellow, created by Carl Martius, was also an injurious color that should be widely prohibited for food and drink. Martius yellow, also known as Manchester, saffron, and golden yellow, was chemically known as dinitro-α-naphthol.[88] However, a sulphonated version of the same dye produced very different results, leading to Weyl to declare naphthol yellow S, or calcium, sodium, or ammonium dinitro-α-naphthol, to be nonpoisonous. Weyl's samples were provided by Gustav Schultz of Berlin.[89] Experiments showed that repeated doses of 0.1 g per kilo of body weight on a young dog, even when injected subcutaneously, did not produce any perceptible disturbances.

> The harmlessness of Naphthol Yellow S is all the more interesting in view of the fact that it differs from the markedly poisonous Martius Yellow merely by one sulphonic group (HSO_3), the introduction of which into a color, renders it soluble. Naturally, one would suppose that a soluble color would prove more poisonous than the insoluble color from which it was produced. Clearly, therefore, we are not in a position to decide as to the poisonous or non-poisonous qualities of any body except from actual research, even when we correctly know the constitutional formula.[90]

It is worth recalling Weyl's earlier comments that this was a market-place where Martius yellow was often sold as napthol yellow S. Mean-while, the presence of salts did not always render a dye safe, as conflicting opinions on the safety of aurantia, or imperial yellow, showed. Aurantia, the ammonium or sodium salt of hexanitrodiphenyl, had been shown by some chemists to be poisonous while others judged it safe.[91] Weyl, un-able to do any experiments on aurantia himself due to a lack of samples, judged the harmlessness of the orange dye to be "doubtful," suggesting that perhaps it acted differently on different people.[92] Despite a sus-tained attempt to organize and control the synthetic dyes through a bat-tery of laboratory tests using the most recent methods for physiological investigation, Weyl and his fellow chemists were repeatedly confounded by the lability of the dyes in variety, name, and chemical composition, and in their tendency to produce contradictory results. Weyl's tests were part of a growing battery of animal (or in vivo) toxicology and physiol-ogy tests being conducted in the nineteenth century. However, it was not until the twentieth century that standardized methods and systematic testing of man-made substances on animals developed.[93] Animal proxy during the late nineteenth century was not a stable method of proof. The Connecticut Agricultural Research Station chemist A. L. Winton ar-gued that tests on animals showing dyes were harmless proved nothing: "Although there is evidence that most of the coal tar dyes are not injuri-ous to some of the lower animals, it is not safe to assume that they are entirely harmless to human beings," claimed Winton, pointing out that the dog, an animal used in most of Weyl's experiments, "has a proverbi-ally strong stomach, and eats with no apparent discomfort many things which would disturb the digestion of a man."[94]

Meanwhile, faced with a growing number of dyes continually enter-ing the market, Weyl admitted that he could not possibly test every azo dye. By the 1880s hundreds of new azo dyes were being prepared, with the German chemical factories of Messrs Lucius & Brüning, Hoechst, BASF and Berlin's Aktien Gesellschaft für Anilin-Fabrikaten (AGFA) leading the field. In 1888, the table of synthetic dyes prepared by the chemists Schultz and Julius detailed 120 azo dyes but Weyl estimated that the number of known dyes totaled "ten times" that figure.[95] In the end, Weyl tested a total of twenty-three azo colors, and determined just two of them, metanil yellow and orange II, to be poisonous. While some of the others, such as Bismarck brown, and fast brown, caused vomit-ing, diarrhea, or protein in the urine of dogs, the small quantities of their use in coloring food and drink was judged by Weyl to be safe.[96]

Some chemists, including the Polish-born, Vienna-based physiological chemist Sigmund Fraenkel, contended that the physiological actions of dyes depended on the group to which they belonged.[97] Fellow Austrian chemist Gustav Schacherl also argued that whole groups of dyes could be identified as being safe or harmful, so that restricting the permitted colors to just a few individual dyes was not necessary. Schacherl, who had studied organic acids with Leopold von Pebal at the Chemical Institute at the Austrian University of Graz, pointed out, however, that far more work needed to be done on testing the physiological impact of different dyes in order to be able to determine the impact of whole classes of dyes accurately. Schacherl recommended permitting all the azo dyes numbered 7 to 393 in the Green tables, except number 86, all the triphenylmethane colors except the hydroxyl derivatives, and all pyronins, oxyketones, and indulins, as well as naphthol yellow S and methylene blue. He advised that all other coal tar colors should be prohibited in food until the group in which they were classified had been determined to be absolutely harmless by physiological testing.[98]

However, other chemists were more skeptical about this approach. The German-born, New York–based chemist Hugo Lieber believed that every individual manufactured dye should be tested by its producer, and sold as a food dye only if proved to be safe by exhaustive physiological tests.

> Although it would be possible to draw quite reliable conclusions as to the advisability of employing certain colors for food products on the basis of their chemical constituency, the mode of their manufacture and of the ingredients used, nevertheless, I think that by far the safest way would be on the one side to force the dealers of colors intended for food products to sell only such colors with which exhaustive and careful physiological tests have been made by experienced and especially impartial and thoroughly reliable people, thereby establishing their harmlessness beyond a doubt. On the other hand, the manufacturers and canners of food products of any description should be forced to purchase and use only those colors which they are sure have been submitted to such careful tests.[99]

Lieber's comments appeared in a book he self-published as promotional material for his company H. Lieber and Co. of New York, an importer and wholesaler of coal tar colors packaged and sold specifically as food colors. Much of the toxicological work included in the book was taken from Weyl's publication, but what appears to be a disinterested public health appeal for pure food colors could equally well be interpreted as a marketing ploy by a dye importer.

G. W. Chlopin, professor of hygiene at the University of Odessa, also disagreed that the toxicity of a dye could be evaluated according to the group in which it was classified. Chlopin judged that tested dyes could be divided into three categories: *poisonous*, those that caused considerable harm to tested animals; *suspicious*, those that caused some form of irritation, vomiting, and diarrhea; and *nonpoisonous*, those that seemed to cause no injury. Chlopin deliberately did not call any dye "harmless," as he claimed that it was impossible to tell whether or not the dyes cause "some finer pathological changes in the organism and functions which could not be detected by simple observation."[100] His investigations of one hundred dyes showed that 30 percent were poisonous and 40 percent were suspicious. However, Chlopin concluded that "it is impossible to say that there is any definite connection between the fact that the dye belongs to a certain chemical group and its action on the animal organism." Any action on an animal organism, he argued, is "determined more by the delicate difference in the internal structure of their molecules than by those differences on which is based, at the present time, the classification of the aromatic dyes."[101] Chlopin made the following "purely empirical generalizations"—that the most poisonous dyes occurred among the yellows and oranges, followed by the blues, browns, and blacks, while he found few poisonous green or violet dyes and only one suspicious red dye. He concluded that coal tar dyes were "as substances foreign to the animal organism, which may influence harmfully the vital functions, even in those cases when they do not possess distinctly poisonous properties," pointing out that many hygienists recommend not using coal tar dyes for food or drink, whether or not they prove poisonous in animal experiments.[102] Chlopin's comments give us an idea of the lability of the dyes and the difficulty chemists had in assessing them. Even if tests showed that the substances were not poisonous, whether the dyes could be considered safe to consume was no straightforward matter. How much caution to exercise and what weight to place on public health or industrial enterprise became a juggling act, where the interests of commerce, the public, and chemists and chemistry had to be balanced.

Experiments using artificial digestive fluids

By the 1890s, chemists were testing the dyes on artificial digestive fluids, introducing another regime of testing into this account. H. A. Weber, professor of agricultural chemistry at the College of Agriculture of the State of Ohio, tested four different coal tar dyes on the digestive enzymes pepsin and pancreatine. While these experiments seemed to pose more questions than they answered—with some of the colors having an effect on pepsin but not pancreatine, while others affected the function of pancreatine but not pepsin—Weber showed that each color interfered with one digestive process or another, casting further doubt on the use of such dyes in food and drink.

> It seems then, so far as these four colors are concerned, that none interfere with both peptic and pancreatic digestion, but that each color interferes seriously with either the one or the other. What the action of other coal tar colors may be, can, of course, not be inferred from this limited number of experiments, but it may safely be said that bodies which have such a decided action in retarding the most important functions of the animal economy, cannot properly have a place in our daily food and drink.[103]

Other chemists were not so sure either about the efficacy of artificial digestion experiments, or about whether the results indicated that synthetic dyes were harmful. Edward Gudeman, a professor of chemistry at the Pennsylvania Museum and School of Industrial Art and at Columbia University, noted that there were drawbacks both with testing dyes on animals and in artificial digestion tests. New York–born Gudeman had studied at Columbia University, then in Berlin and Göttingen. Later, after 1900, he would go on to work in private practice as a consulting chemist and expert witness. Gudeman raised two important, but different, problems associated with testing dyes on animals and in vitro. The main objection with feeding dyes to animals, he argued, was that the quantities introduced into the system were much greater than what was normally to be found in food products and, as a result, any ill effects were likely to be due to this excess. Doses given to animals that were similar to the amounts found in food, even when administered over a long period, seemed to have no more effect than "harmless condiments" such as spices, sugar, salt, vinegar, or alcohol.[104] The main question hanging over artificial digestion tests, according to Gudeman, was whether they could re-create the complex processes involved in digestion. "The strongest objection urged against the artificial digestion experiment," he wrote, "is that such tests do not supplicate the complex action of the organized system and therefore are not at all conclusive as to what the effect of a color, preservative or condiment will be when consumed in a food product or with a food product."[105]

Gudeman conducted numerous experiments to test how dyes reacted with pepsin and pancreatine, using uncoagulated egg albumen, potato starch, and high-proteid maize gluten flour to replicate food. According to Gudeman, "not a single one of the colors tested in proportion of 1 part to 1600 parts of egg albumen, was found to affect the peptic nor pancreatic digestion *in vitro*." The tests were not just applied to synthetic dyes (Hofmann's violet 3B, Paris violet, Bismarck brown, eosin, fuchsin, synthetic indigo, Congo red, naphthol yellow, ponceau 2R, amaranth red, turmerine yellow), but also to mineral (Prussian blue, ultramarine, burnt sienna, chrome yellow, and iron oxide), animal (cochineal), and vegetable (turmeric, annatto, cladonal red) colors used in foods. The results indicated that of the colors used, only ultramarine, burnt sienna, chrome yellow, and ponceau 2R affected artificial digestion with pepsin when used in quantities of 1 part of the color or less to 400 parts of the food. The results also indicated that synthetic colors were less active than both mineral- and animal-based colors, and no more active than the vegetable colors. Both vegetable and synthetic colors were directly digested by pepsin and by pancreatine, with little difference between the two. As a result of the experiments, Gudeman concluded that synthetic colors were less active in retarding the action of digestive fluids than any other class of colors, partly because of the fact that their coloring power was so strong that only minute amounts of the synthetic dyes were required to produce the effect equivalent effect to that of large amounts of natural colors.[106]

A lack of consensus

There was little consensus among nineteenth-century scientists about which dyes were safe or even about how to test the dyes, a situation not helped by the lack of consistency in naming and manufacturing dyes and the trade secrecy surrounding their production and supply.[107] Unlike the dyes themselves, increasingly manufactured in large state-of-the-art industrial research laboratories, the tests to detect their presence in food and their toxicity were conducted in a variety of settings, from institutional and educational facilities in university departments to state-sponsored research facilities to private laboratories, often located in chemists' homes and shops.

By the turn of the century, legislators across the United States and Europe were struggling to address mounting public and political concern surrounding the use of chemical dyes in food. Despite the application of a range of testing regimes applied to the issue, little scientific consensus was reached concerning the safety of artificial dyes as food colors. In 1906, Hesse described the diverse range of positions taken by chemists regarding the toxicology of synthetic dyes, after reviewing contemporary scientific literature as part of the evidence gathering for US regulations on the use of coal tar dyes in food. The lack of consensus in the scientific community found by Hesse was notable.[108] For example, German ophthalmologist Jakob Stilling, who used aniline dyes in his treatment of eyes, argued that aniline colors were harmless, and claimed that the deaths of the subjects in animal experiments "prove nothing." The German chemist Karl Lehmann, professor of hygiene at the University of Würzburg, meanwhile, warned that the harmlessness of the aniline and azo dyers was not yet proven, despite conceding that manufacturing improvements had reduced the toxicity of synthetic dyes by, for example, removing arsenic.[109]

Other chemists argued that every dye should be treated as suspicious until it had been tested by recognized physiological experiments.[110] However, with hundreds of dyestuffs on the market and more appearing every year, testing every dye and reaching a comprehensive consensus about each one was judged by many chemists to be impossible. The fact that analytical capabilities at the time fell behind synthetic capabilities was also an issue. What modicum of consensus was achieved centered on the extensive research conducted by Weyl, whose work was cited widely by other chemists in journals and reports. For example, the Society of Swiss Analytical Chemists recommended that picric acid, dinitrocresol, Martius yellow, aurantia, orange II, metanil yellow, safranin, and methylene blue, all the colors identified by Weyl as harmful, be prohibited for use in food and drink.[111]

Despite a lack of complete consensus, the repeated publication of lists of harmful dyes led many to the implicit assumption that those not on the lists were safe, even though tests had not been done on most of the hundreds of coal tar dyes on the market. This misleading situation was highlighted in an 1896 article in *Food and Sanitation*, referring to a bulletin issued by the dairy and food commissioner of Pennsylvania that identified seven coal tar colors possessing "marked poisonous properties." Almost certainly taken from Weyl's results, these poisonous dyes were identified as Martius yellow; dinitrocresol; picric acid; metanil yellow; orange II; dinitroso-resorinol; and naphthol green B. Aurantia, meanwhile, was classed as "suspicious." The article observed that coal tar colors were extensively used to color foods, confections, liquors, and soft drinks and that, while many countries had introduced, or were in the process of introducing, legislation to prevent coloring by dangerous dyes, in most cases "all that is attempted is to prohibit the use of those colouring matters known to be injurious to health." Meanwhile, new coloring matters were continually appearing in the marketplace, noted the author, recommending that "in order to prevent manufacturers from experimenting upon the public, the use of such substances should be prohibited until their influence upon the human organism has been investigated."[112]

To ban all coal tar dyes in food, after food producers and consumers had become accustomed to their use over several decades, would have been a politically and commercially difficult approach. Instead, governments took a variety of actions, including banning certain dyes that were known to be toxic or recommending a restricted list of harmless dyes for use in food. Approaches varied according to different national cultures and philosophies, from countries adopting precaution as a principle of public policy to those favoring a laissez-faire, buyer-beware approach. Regulators turned to chemists to choose which dyes to ban or recommend, and how to regularly update such lists reliably. With no consensus or standardized testing for detecting and assessing the dyes, chemists and regulators fell back on widely publicized results such as Weyl's. However, as new dyes continued to be manufactured and new tests devised, there was little consistency over which test to use.

Chemists struggled to know and understand these new substances that came to be used in varied and often unintended ways, and whose properties, by their very nature, changed when combined with other chemicals. Devising tests to detect and assess hundreds of new, and constantly appearing, substances was a formidable task, made more difficult both because most of the substances were new to the testers, and because their number was constantly increasing. Reaching any kind of consensus among chemists, from different countries and disciplinary backgrounds, over which tests were most appropriate or effective, or over how to standardize tests and their results, proved almost impossible.

While the analytical skills of nineteenth-century chemists led to their appointment as public guardians of the food supply, their analytical skills always lagged behind the synthetical skills of the chemical manufacturers, who produced new dyes at an accelerating rate. Regulators and public analysts were faced with the difficulties associated with creating public policy over substances that were constantly changing and proliferating. In this, the nature of the dyes as objects of knowledge and the capacity of chemists to test for them would be critical. One can test for and ban known substances but not for new, unknown ones, or for the endless possibilities created by countless interractions between different chemicals, a situation that has continued to perplex policymakers throughout the twentieth and twenty-first centuries.

Without the ability to reliably test for the ever-increasing numbers of synthesized dyes appearing in the marketplace, the ability of chemists to mediate on the use of the new chemicals in food depended on their authority and relationships with others and varied from country to country.

Experiments for detecting the toxicity of the new dyes, as well as the experiments outlined to detect their use in food, demonstrate clearly the variety of cultural and technological approaches marshalled by the scientific community in an attempt to solve the problem of the paradoxical dyes, at once so colorful and so elusive. Negotiations over how synthetic dyes were to be understood, tested, manipulated, and used as substances were complex, involving not only chemists but also politicians, producers, and consumers. As Bodewitz and others have argued, "the stabilization of technologies and artefacts is bound up with their adoption by relevant social groups as an acceptable solution to their problems."[113] One of the most problematic aspects of evaluating these new substances was the inability to identify and quantify them in standardized terms.

As historians of science have pointed out for other disciplines, the adoption of an agreed-upon classification system is as much political, social, and commercial as it is scientific. While chemists agreed on the need for a uniform classification and nomenclature of dyes, the task was hugely complex, not least because of the lack of standardization in the notation of chemical formulae during this period. During the second half of the nineteenth century, scientists sought to standardize measurements such as time, weight, length, and electrical units to create a common language and set of values in an increasingly transnational environment. Through international conferences, journals, and other forms of national and transnational scientific dialogue, scientists established uniform methods of measuring, unitizing, nomenclature, and classification across the natural sciences and technology in this period. Many of these standardization projects have been well documented and shown to be far from straightforward tasks, involving a complex process of negotiation.[114] However, the task became ever more complex as scientific processes and products became more commercialized. While units of electrical measurement were agreed on and standardized in the nineteenth century, it was not until well into the twentieth century that uniform standards were applied to *consumer* electrical supply systems and appliances. Trevor Pinch argues that reaching consensus becomes more difficult the broader the decision-making group and social context become.[115] Standardization becomes a far more complex situation to resolve within a competitive marketplace, where the negotiation includes manufacturers, retailers, the public, and politicians as well as scientists.[116] Tests involving the body and attempts to calibrate bodily senses such as taste and sight were particularly complex and compelling for nineteenth-century scientists.[117]

As Simon Schaffer observes, "quantification is not a self-evident nor inevitable process in a science's history, but possesses a remarkable cultural history of its own."[118] In the case of chemical dyes, new strategies had to be devised, with chemists from a variety of backgrounds and disciplines and with different aims and objectives, drawing on a diverse range of repertoires of testing. According to Schaffer, "the values which experimenters measure are the result of value-laden choices," and the phenomena and criteria that are measured depend on the technologies adopted and available.[119]

Schaffer and other scholars in the history and sociology of science, such as Harry Collins, have demonstrated the complexities of achieving consensus in the creation of universally agreed-upon tests, stressing the social negotiation involved in determining which experiments and "facts" are applicable. The situation in assessing chemical dyes was particularly complex. These were novel and undetectible substances being created by chemists, themselves unable to agree on their nomenclature or chemical composition, and which were rapidly permeating invisibly throughout society, to be detected, analyzed, and assessed by scientists from different disciplines with diverse traditions and objectives. As Collins explains in his theory of experimenters' regress, "usually successful practice of an experimental skill is evident in a successful outcome to an experiment, but where the detection of a novel phenomenon is in question, it is not clear what should count as a 'successful outcome'— detection or non-detection of the phenomenon."[120] This was a situation with uncertain facts and a widespread and disparate group of parties seeking agreement. Moreover, even when consensus was reached on which type and form of experiment should be used, experimental replication was often problematic. The transfer of experimental practice from person to person, application to application, and setting to setting is seldom straightforward and can be complicated by apparatus, conditions, and materials of variable consistency.[121]

By examining how chemists understood the new synthetic dyes as objects of inquiry, we can clearly see the successful interplay of experimental practices and theoretical knowledge, as well as the continued productive partnership of analysis and synthesis in chemistry.[122] This challenges earlier historiography of synthetic and analytical chemistry, based on accounts such as those by J. R. Partington and Adolf von Baeyer, that stressed a growing divide between synthetic and analytical chemistry and suggested that the synthetic chemistry of the "Second Industrial Revolution" was secure and based on universally agreed-upon structural theory.[123] These previous accounts implied that the eighteenth-century coupling of analysis and synthesis in chemistry, based on the experimental and hands-on practice of decomposition and reconstitution of substances, gave way to a chemistry based on theoretical understanding and knowledge, following the development of synthesis and structural theory by chemists such as Kekulé. Scholars such as Catherine M. Jackson, Ursula Klein, and Alan Rocke, however, present a more nuanced and complex picture, arguing that huge uncertainties remained in the understanding of molecular and chemical structure throughout this period and that chemists deployed a variety of theories and experimental methods and techniques in interpreting and manipulating organic chemistry.[124] As we have seen, chemists employed theory and practice from across a wide range of chemical traditions and developments, combining both synthesis and analysis, in their attempts to detect the use of coal tar dyes in food and to understand and assess the impact they had on those consuming them.

5

The appointment of public food analysts in Britain

Municipal and state chemists, or public analysts as they were generally called in Britain, were a nineteenth-century creation and first appeared in Britain in the 1870s. The fact that Britain was one of the first industrialized countries to implement comprehensive food regulation and to legislate for the appointment of public analysts to help safeguard the nation's food but one of the last to legislate on the use of coal tar dyes and other synthetic chemical additives in food makes it a particularly interesting case study. Learning more about how this new type of professional chemist came into being and some of the early struggles that they encountered in establishing any authority or credence helps explain how they responded to the introduction of synthetic chemicals into food.

As already discussed, analytical chemists in Britain, as well as other countries, were highly critical of the use of minerals and metals to color foods. Indeed, the publicity surrounding poisonous metallic coloring in confectionery and other foodstuffs helped to promote measures to protect the public and to establish the public analyst's role as a gatekeeper of food safety. But while concern about artificial food coloring from mineral substances, such as lead and arsenic, played a crucial role in early attempts to regulate food production, new organic chemical colorings were entering the food supply at a time when the public analysts were not secure enough either in their status or in their chemical knowledge and testing techniques to question these new "wonders of science."

Struggles surrounding professional jurisdiction and the search for public authority, particularly in analytical chemistry, were a key feature of health debates in the nineteenth century. Questions of managing public health, nutrition, and food and drink adulteration were problematic and contested issues from the outset, with health problems associated with milk, water, meat, and other comestibles becoming sites for challenging professional authority.[1] Chemical dyes entered the food supply in the midst of these struggles.

While food adulteration had been of growing concern from the eighteenth century, it was the increasing use of toxic mineral colorings that finally prompted a parliamentary select committee hearing into food adulteration in 1855 in the House of Commons. Experts who addressed the committee included Wakley; Henry Letheby; Hassall; John Postgate; Alphonse Normandy, a London-based, French-born, and German-educated consultant analytical chemist; Robert Thomson, professor of chemistry at St Thomas's Hospital; Theophilus Redwood, a founding member of the Pharmaceutical Society; Robert Warrington of the Society of Apothecaries; and George Phillips, principal of the chemical laboratory of the Board of the Inland Revenue.

Several of the chemists, especially those with pharmaceutical experience, pointed out to the committee that substances could be impure but nevertheless suitable for the purposes for which they were intended, and harmless. They also sought to differentiate between intentional and unintentional adulteration, as well as between adulteration that sought to deceive and defraud the public for monetary gain and that which could cause physical harm. Redwood, Letheby, and Phillips all claimed that harmful adulteration was not as extensive as Hassall's revelations implied.[2] Ironically, however, some food manufacturers who addressed the committee, most notably Thomas Blackwell, whose company produced pickles and preserves, claimed adulteration was rife and needed to be controlled. While manufacturers and retailers generally complained about the intrusion of government regulation, the larger firms believed greater restrictions and prosecutions relating to food adulteration would hit their smaller competitors harder, giving larger producers a competitive advantage. Blackwell admitted that his company used copper salts to make preserved vegetables appear greener and colored meat sauces red using iron compounds, unaware that such additives were considered disagreeable.[3] Other manufacturers linked adulteration to foreign imports or argued that trade practices that could be deemed as adulteration were, in fact, used to meet consumer expectations. The use of additional ingredients allowed food manufacturers to produce a wider range of food products at a cheaper price, company executives argued.[4]

The committee's report led to the first British food adulteration act in 1860. However, this legislation, like subsequent food legislation in Britain during the nineteenth century, failed to define adulteration and made proof of knowledge of wrongdoing on the part of the accused mandatory for prosecution. As a result, prosecutions under the act focused more on the deliberate misrepresentation of food, helping to normalize trade practices involving the introduction of additional ingredients into existing food products.[5] The inadequacy of the 1860 act in combatting food adulteration led to further food legislation in 1872. Once again, the 1872 act failed to define adulteration, concentrating instead on added ingredients that could be proved harmful and whether the manufacturer or retailer disclosed additional ingredients to the consumer. Although the 1872 act recommended the appointment of public analysts, its permissive nature meant that by 1874, only 26 out of 171 boroughs and 34 out of 54 counties had appointed public analysts. Continuing worries about adulteration, together with lobbying from both manufacturers and retailers, who felt they were being unfairly prosecuted, and chemists, who felt that public analysts should be given more power and better training, led to a further select committee hearing in 1874. The Read Select Committee was chaired by Clare Sewell Read, parliamentary secretary to the local government board and the member of Parliament for the farming constituency of South Norfolk. The committee of MPs heard fifty-seven witnesses and discussed, among other things, the roles of public analysts and the government's excise chemists as professional gatekeepers of food purity. The committee's report justified the addition of harmless additional matter into food such as coloring in butter or cheese as long as it was harmless and not added to deceive the consumer. The committee also agreed with food manufacturers that admixtures, as long as they were disclosed, were necessary to meet consumer expectations and to provide affordable food products.[6]

The 1875 Sale of Food and Drugs Act, which sought to enact the committee's recommendations, is widely regarded as the basis of modern British food adulteration legislation. However, this act too omitted any definition of adulteration. While it introduced strict liability in that the introduction of harmful adulterants was sufficient for a prosecution to be made under the act, the prosecution still needed to prove intent on the part of the supplier. Meanwhile, the 1875 act effectively legitimized the use of additional substances into food by allowing patented and labeled admixtures of ingredients provided that they were not harmful and enabled the production of food "in a state fit for carriage or consumption." As a result of the 1875 act, the state helped to create the legalization of new food ingredients and practices.[7]

Sébastien Rioux argues that the regulation of food standards in nineteenth-century Britain helped advance "capitalist food production and the rise of legal adulteration."[8] The present book shows that chemists played a major part in both the creation and the legitimization of synthetic chemicals in food production, leading to a degree of legal adulteration that the nineteenth-century state and public could have never foreseen.

One of the major changes introduced in the 1875 act was the mandatory appointment of public analysts. Historians have demonstrated the extent to which "scientific expertise" influenced public health legislation and its enforcement during the second half of the nineteenth century.[9] Public health legislation was part of a growth of government legislation during a period in which Victorian government moved from laissez-faire free trade regimes through to permissive legislation and then to increasing enforcement of laws by local authorities and the central government.[10] The relationship between public analysts and new scientifically devised substances and the ways in which they were integrated into our food supply sheds further light on the development of government legislation, risk management, and the involvement of expertise as well as the legitimization of new food additives.

Formation of the Society of Public Analysts

The Society of Public Analysts was formed in 1874, in response to the Read Select Committee and criticisms about the inexperience of public analysts and their inconsistent decisions, as well as a lack of consensus over what constituted adulteration. Dr. Theophilus Redwood, professor of chemistry at the Pharmaceutical Society and public analyst for the County of Middlesex, who took the chair for its first meeting on August 7, 1874, described the aims of the meeting as "first, the refutation of unjust imputations; secondly, the repudiation of proposed measures of interference with our professional position and independence; and thirdly, the formation of an association having for its objects the promotion of mutual assistance and co-operation among public analysts."[11]

Such words, as well as correspondence and reports of the SPA's meetings, reveal the degree to which these new public appointees felt the need to justify their position and demonstrate their expertise. Understanding the context in which public analysts operated during the nineteenth century, and exploring their ongoing disputes with fellow men of science, including doctors and other chemists, as well as among themselves, helps explain the hesitant position the analysts were to take regarding the emergence of new chemical substances incorporated into food products.

Public analysts were one of several new categories of scientists attempting to establish their position among an unfolding hierarchy of new professional "experts." Disputes surrounding methodologies, pay, and status reflected the divide between the prominent chemists of the day, many of whom had academic posts, such as Edward Frankland, and the practicing professional chemists, such as public analyst James Wanklyn, who were anxious to assert their authority and status through their own affiliations and professional associations.[12] The public analysts clearly felt that they were an undervalued group of scientists forced to strive for recognition and financial recompense. In 1880, John Muter described their work as being carried out "not by wealthy *dilettante* or men receiving grants from research funds, but by persons daily and actively engaged in carrying out the complex and tedious duties thrown upon them by an Act of Parliament."[13] As historians of science such as Ian Burney, Graeme Gooday, and Christopher Hamlin have shown, the negotiation of expertise and authority among "men of science" and others was complex and contested during the latter half of the nineteenth century.[14] The tortuous and lengthy journey chemical analysts made to establish themselves as recognized professional scientists was commented on by Dublin's first public analyst, Sir Charles Cameron, in his president's speech to the Society of Public Analysts in 1894. Cameron observed that within his lifetime, chemical analysis had broadened out of the laboratories of professors and lecturers into the commercial and private laboratories of practicing chemists and companies, thanks largely to the agricultural fertilizer industry and the food adulteration debate. In twenty years, he noted, the status, ability, and salaries of public analysts had increased significantly from the low regard and poor pay they faced in the immediate aftermath of the 1872 Adulteration Act.[15] However, public analysts continued to complain until well into the twentieth century that they were underresourced and lacked government support in their efforts to protect the public from fraudulent adulteration.[16]

Throughout this period there was constant quibbling among chemists and physicians as to whose skills and expertise were best suited to defeat food adulteration. Even before legislation had introduced the position of public analyst, the publication of Hassall's report on adulteration led to a bitter and lengthy priority dispute as to how much credit should be assigned to the doctors campaigning against adulteration like Wakley and Postgate, the microscopic skills of Hassall, and the analytical skills of Henry Letheby, public analyst and medical officer for the City of London.[17]

In the early decades of food analysis, public analysts in many local authorities were shared appointments with medical officers, resulting in the posts being filled by medically trained men with little chemical expertise and subsequent accusations of poor chemical analysis. Even when the post of public analyst was not shared with the medical officer post, the skill set required of the candidate was open to negotiation. Part of the dispute as to whether chemical or medical expertise was most appropriate for a public analyst centered on whether the primary concern of adulteration was considered a public health issue or an issue of economic fraud, a matter of much debate. The main focus of the 1860 and 1872 Sale of Food and Drugs Acts was the use of poisonous substances in food, such as toxic metallic dyes, and the 1872 act described medical, microscopical, and chemical knowledge as important skills for dealing with the issue of adulteration. The 1875 act subsequently removed the requirement for medical knowledge and focused on the concept of economic fraud, in which the purchaser might be prejudiced if sold a good "not of the nature, quality, and substance demanded." As a result, economic harm, rather than medical harm, became "central to British understanding of adulteration for the next 100 years," according to Atkins.[18] The emphasis on food fraud over food safety helped secure chemical analysis as the foremost tool in food monitoring. However, increasing concern over food hygiene and the spread of diseases through food and milk, particularly as the food supply chain lengthened, meant that physicians and chemists continued to squabble for decades over who was better qualified to oversee the public food supply. In 1898, the *British Medical Journal* complained that a board of reference proposed to oversee the enforcement of food and drug legislation comprised five chemists but only one "medical man." "Considering that the physiological action of impurities, both in food and drugs, is the real ground of legislation in the matter, this seems hardly fair to the public whose constitutions are to be protected from disease. The representation of the analysts is essential, but of the biologists not less so," the journal argued.[19] The increasing use of chemical additives, in the form of both preservatives and colorings, to increase the shelf life of food and the desirability and attractiveness of processed food, complicated the situation further. Such chemicals were considered an improvement to food quality and availability while, at the same time, regarded as a means for deceiving the consumer as to food quality and freshness.

The growing, but still small, band of publicly appointed chemists, on whom the new adulteration legislation depended, decided to set up their own association to fight their corner and promote their cause, forming the Society of Public Analysts and publishing their own journal, the *Analyst*, from 1876. Although the British Association for the Advancement of Science had a chemical section from its formation in 1831, it was not until 1841, when the Chemical Society of London was formed, that chemists in Britain had an association of their own, the same year in which the Pharmaceutical Society of Great Britain was created for "chemists and druggists." As Chirnside and Hamence point out, many members of the Chemical Society, even in the 1870s, were still "amateur and theoretical chemists" or "interested clergymen."[20] The creation of separate societies and institutions contributed to a growing distinction between different sectors of the chemical milieu, including academic, consulting, amateur, and retailing and wholesaling chemists and pharmacists.[21] This was a time when professional chemists were attempting to define who they were and to differentiate themselves from retail druggists and other practical chemists. Even the retail chemists were split between pharmacists selling patented drugs and those selling generic products, a similar situation to that in Germany.[22]

Practicing chemists during the nineteenth century, that is those not chiefly employed in academia, in industry, or as dispensing chemists, generally earned their living in private practice as consulting and analytical chemists. This often meant working for many different clients in diverse sectors of commerce. As William Brock has shown for the case of William Crookes and other practicing chemists of the period, success involved taking jobs whenever and wherever available, alongside self-promotion and public endorsement of the chemical profession in general. Crookes, who founded and edited the journal *Chemical News*, was one of many chemists to use the burgeoning Victorian media to promote chemistry.[23] Adulteration proved to be the perfect fodder in a flourishing symbiotic relationship between the press and chemistry. For example, Charles Cameron, the editor of the *North British Daily Mail* and a doctor with experience of chemical analysis, used his publication and the services of a young analytical chemist to expose the adulteration of whisky in Glasgow in the late nineteenth century.[24] Cameron, not a relation of Dublin's Charles Cameron,[25] became a member of Parliament in 1874, and introduced a bill to amend the Food and Drugs Act in 1894. Although the bill was withdrawn, a select committee was set up to examine food adulteration and the existing legislation, leading to an amended act, passed in 1899, which required proof of qualifications for the position of public analyst.

A review of the *Analyst* from its foundation to the late 1800s clearly shows the extent to which the public analysts were fighting to establish their credentials and expertise in a new but already crowded field of professional men of science.

In its early years, the SPA felt under attack from all sides, including from the public, who complained that taxes were being paid to public analysts who did little to prevent adulteration. The *Analyst* suggested that the problem was that the new 1872 Adulteration of Food and Drugs Act put the onus on complaining members of the public to prosecute rather than the local authority, whose job was purely to analyze. As a result, the SPA urged the public to do their duty and prosecute, promising "that if they will awake to a sense of their duty, every aid will be given them by the much abused Public Analysts."[26] Consumers were not passive agents and were well aware of the issues surrounding adulteration, and the SPA was presenting itself as a mouthpiece for the consumers.[27]

However, it was disputes between the public analysts and other men of science, notably and paradoxically fellow chemists, that attracted much more comment in the pages of the *Analyst*.[28] The place of the chemist in adulteration cases was both ambiguous and contested. Many food and drink ingredients came to be viewed as adulterants over time, including licorice, ginger, turmeric, alum, copperas, and sulphuric acid. Such products were supplied to the Victorian food and drink trade by chemists and druggists.[29]

The profession of chemist, and scientists in general, was undergoing a rapid transformation and the experience and training of many of its members spanned across many areas that would be perceived as separate disciplines only in the following century. Chemists operated both in academia and in commerce, working in both public and private practice. They invariably knew each other and indeed had often been trained by the same mentors.[30] This, in many ways, made their disputes more complex and imperative as they strove to assert their own opinions about the best way to secure their professional interests and sought different allegiances to help them.[31]

Professional disputes and grievances were the source of much mudslinging in the growing number of professional and trade publications. Particularly antagonistic were the exchanges between the *Analyst* and the *Chemist and Druggist*, perhaps because many of the suppliers of the adulterants being detected by the newly appointed public analysts were the wholesaler and retailer readers of the *Chemist and Druggist*. Jibes complaining that the *Chemist and Druggist* and other trade journals had no understanding of chemistry, and disparaging references to druggists and pharmacists' motives and lack of ability, continued to make their appearance in the *Analyst*. Frequently critical of other journals' obstructions of the public analysts in their work, the *Analyst* also accused trade journals of acting for the benefit of their advertisers. Conflict between the public analysts and the grocery trade was also apparent in continual spats between the *Grocer* and the *Analyst*. The *Analyst* urged the *Grocer* to desist from calling analysts incompetent and to work with analysts to stamp out adulteration.[32]

Defining what constituted adulteration was a matter of ongoing dispute between the food industry and the public analysts. Food manufacturers often argued that if an added ingredient was not injurious to health, then it should not be considered an adulteration, particularly if it made the food more palatable and acceptable to the consumer. Analysts, however, took a less commercial attitude to food adulteration, arguing, for example, that if food was being sold as "cocoa," it should not contain substances other than cocoa. Discriminating between "commercial" and "analytical" purity was a basis for many court cases, particularly when food items were sold as "pure" but contained additional nonharmful ingredients often used as part of the manufacturing process.[33]

Speeches, reports, and articles in the *Analyst* clearly show how public analysts were promoting themselves as the impartial gatekeepers of food purity with the chemical truth of the laboratory as their rational weapon in the mire of the Victorian free market economy.[34] Not only were analysts fighting the fraudsters in the market, they also had to assert their authority amongst their peers. As Hamlin has shown in his study of nineteenth-century water analysis, there was considerable anxiety among practicing and consulting chemists over the very public and acrimonious debates between prominent chemists such as Hassall, Letheby, Wanklyn, Frankland, and others over methodologies and contradictory results in chemical and microscopical analysis.[35]

Public analysts felt they were under attack from the outset, and were especially angry about the evidence given by the chemist Augustus Voelcker before the Read Committee in June 1874. Voelcker, who was a member of the Royal Agricultural Society and had his own private practice, said that he "doubted there were more than a dozen chemists in the whole country capable of carrying out the duties of public analyst properly," and was unsparing in his criticism of the new body of public analysts:

> a good deal of mischief has been done by the so-called analyses, and the food analysts have been the greatest enemies to the Food Act. . . . Many are very incompetent analysts; they have not had sufficient chemical training, nor any experience in analysis, and therefore their statements are sometimes of a very flippant and unwarranted character, which has done a great deal of harm in rendering the Act contemptible in the eyes of practical men.[36]

Voelcker was born and educated in Germany, where chemical training, particularly in organic chemistry, was notably more advanced. He had begun work as a pharmacist's assistant in Frankfurt, studying under Friedrich Wöhler at the University of Göttingen, Liebig at Giessen, and finally Gerardus Johannes Mulder in Utrecht. A connection with Germany was not unusual among SPA members, since several public analysts, like Otto Hehner, were German, or had traveled to Germany at some point in their training, such as James Wanklyn and August Dupré. In 1863 Voelcker set up a private consulting practice, advising on a variety of matters including agriculture, sewage, water, and gas as well as undertaking adulteration analysis on agricultural fertilizers and feedstuffs. Despite his earlier criticism of the state of chemical analysis in Britain, Voelcker went on to become president of the SPA himself in 1901 as well as training future analysts, including Bernard Dyer, who would become the SPA's president in 1897. Voelcker was keen to see the professional status and standing of public analysts improved through education. However, his comments to the Read Committee, particularly his recommendation that the government excise chemists be appointed as arbiters in adulteration disputes, antagonized fellow public analysts such as Alfred Allen, the public analyst for Sheffield, and James Wanklyn, the public analyst for Buckingham and Shrewsbury, among other boroughs. Professional rivalry between public analysts such as Allen and Wanklyn and the government chemists was acrimonious and lasted for several decades.[37]

Somerset House: the Government Laboratory

The Government Excise Laboratory was the state's official laboratory, based in Somerset House from 1852 to 1897. Originally named the

Board of Inland Revenue Chemical Laboratory (1842–1894), it subsequently became known as the Government Laboratory (1894–1911)
and then the Department of the Government Chemist (1911–1959). For
much of the nineteenth century it was simply described as Somerset
House.[38] One of the earliest official state laboratories in the world, it
was originally set up by the Excise Office in the wake of the 1842 Pure
Tobacco Act to ensure that the state was not cheated out of taxes due
on tobacco and other goods, and to reduce the costs incurred in having
samples analyzed by external chemists. The first principal chemist was
George Phillips, who had begun work as an excise officer in 1826. Phillips, self-educated in chemistry and microscopy, remained with the laboratory until his retirement in 1874, when James Bell, his deputy from
1867, succeeded him. Like its two principals, many of the laboratory's
staff were originally excise officers who had spent most of their careers
with the Excise Department.[39]

From its early days, the laboratory was criticized by academic and
consulting chemists who were employed by manufacturers and retailers to argue their position against the Excise Department. Thus, it is
no surprise that external chemists sought to create doubt surrounding
the government officers' chemical expertise. The department's chemists
were also criticized by Hassall and others for their incompetence and
lack of experience in analyzing the adulteration of food, as a result of
their focus on taxable products such as tobacco, tea, and coffee.[40]

As food adulteration became a growing social issue, the work of the
Excise Laboratory grew fast, with the annual number of samples analyzed by the Excise Department rising from 100 in 1842 to over 9,500
in 1859.[41] By this time, the laboratory was operating as a separate department within the Inland Revenue. Initially, the department employed
assistants to work as chemists on secondment from their usual work
as excise officers while continuing to use outside analysts. The government officers were given chemical training at University College, London (where the Board of Excise chairman, John Wood, was a member of
the university council). By the late 1850s, in order to save money both
in training costs and by reducing the amount of work going to external analysts, it was decided to train the department's chemists in-house
under Phillips. Phillips's students were examined initially by Hofmann
of the Royal School of Mines and, from 1866, by Hofmann's successor,
Edward Frankland.[42]

As career civil servants, with limited interaction with the close-knit
but growing community of professional consulting chemists, and with
a history of, and reputation for, primarily protecting the country's rev-

enues rather than its health and the quality of its nontaxable food supplies, it is perhaps not surprising that the Inland Revenue chemists attracted critical comment relating to their alleged lack of expertise in general food analysis. However, while the staff may have been excise officers first, laboratory notebooks and records demonstrate their chemical ability, and most of the government chemists would specifically have asked to be moved to the laboratory because of their interest in chemical analysis. Indeed, many of the Excise students, such as Phillip's assistant George Kay and James Bell, won chemistry prizes at University College.[43] Their credentials as chemists were thus well-established.

Following recommendation by the Read Parliamentary Committee, the Government Laboratory was appointed to test for disputed samples of alleged adulterated food and drink sent to magistrates' courts. Although the actual numbers involved were small, the cases became the focus of increasingly hostile disputes between the government chemists and the public analysts, threatening the authority and perceived expertise of both. The public analysts struck out at the government chemists from the outset, with Wanklyn's objection to the Read Committee's recommendation that Somerset House act as a court of appeal for disputed cases of adulteration:

> The Excise Laboratory does not enjoy a high repute among chemists . . . I do not think that any good could arise from having official referees in those dispute cases. I have had some experience of the kind myself, and I can promise you that if you were to elect any such body, and give them powers of this description, you would be giving them a license to blunder and to disregard the general views of the chemists.[44]

Wanklyn was an exceptionally outspoken public analyst, not afraid to confront the establishment. The SPA's president, Bernard Dyer, described him as having "a spirit of pugnacity which seemed almost inevitably to involve him sooner or later in personal quarrels."[45] Wanklyn had already had a public falling out with the eminent chemist Edward Frankland over water analysis and was subsequently to resign from the SPA over another dispute. Hammond and Egan suggest that Wanklyn's prior dispute with Frankland perhaps fueled his intense animosity towards Somerset House.[46] It was also not unusual for analytical chemists to find themselves on opposite sides of a courtroom as expert witnesses. Frankland and Wanklyn found themselves in such a situation in 1864 when the Lyons-based dye manufacturer Renard Frères sued

Manchester-based dye manufacturer Hugo Levenstein for infringing its patent rights to fuchsine (magenta or rosaniline).[47]

Like Frankland and some other English chemists of the time, Wanklyn had studied at Owens College in Manchester followed by further training in Germany, at Robert Bunsen's Marburg laboratory. With Frankland's help, he had secured a job with Lyon Playfair before becoming a lecturer at the London Institution from 1863 to 1870, and then juggled work as a public analyst to several local authorities as well as working as an analyst in private practice with a lectureship at St. George's Hospital.[48] Wanklyn's educational background and working mix of private and public consulting and part-time lecturing were typical of his peers in the SPA.

The disputes between government chemists and public analysts became the subject of comment and ridicule in both the trade and public press. A correspondent in *Food, Drugs and Drink* referred to the struggles "between the conscientious public analyst and that scientific-analytical-laughing-stock—Somerset House."[49] Facts, methodologies, and interpretations were disputed not only in the press but also in parliamentary hearings and in courtrooms, with companies, the government and prosecution and defense lawyers all employing individual analysts to support their case. As one judge noted in 1866, after listening to conflicting evidence from three chemists, such witnesses were always able to create a scientific argument according to who was employing them with a clear conscience, due to the inherent uncertainty of the science.[50]

Public analysts, like other professional scientists, were no strangers to Victorian courtrooms, often being employed as expert witnesses.[51] Conflicting evidence presented by chemists in court became the subject of much controversy in the second half of the nineteenth century, producing a lengthy and heated debate as to whether scientists should be deployed as expert witnesses.[52] An 1885 editorial in *Nature* criticized William Odling, president of the Institute of Chemistry, for promoting a professionalization and commercialization of science whereby chemists were paid to be consultants to businesses as well as public authorities and appeared as witnesses for both prosecution and defense in court.[53] Indeed, by the mid-1880s, expressions such as "liars, great liars, and scientific witnesses" were widely circulating in legal and journalistic communities, and in 1891 *The Spectator* scathingly censured men of science for giving up the pursuit of pure truth and knowledge for the "fat plains and valleys of the material world."[54] Tal Golan and Graeme Gooday agree that scientific "experts" arguing in courtrooms became the subject of some ridicule during the latter nineteenth century and impaired the efforts of men of science to portray themselves as impartial arbiters of the truth.[55] However, Odling and William Crookes, a consulting chemist and editor of *Chemical News*, claimed that chemists and other scientists played valuable roles as consultants and witnesses, and argued that a lack of standardization and different analytical processes, as well as the fact that all cases were different, led to conflicting views and evidence.[56]

The courts became another arena, alongside the laboratory, the Parliament, the marketplace, and the home, in which the quality of food was determined.[57] The difficulty for the courts, and the expert witnesses, was the lack of consensus over what was "normal" and "natural" food and what was "adulterated." Public analysts felt the need to have defined parameters and standards in order to distinguish legally between "normal" and "adulterated" food as they increasingly found themselves having to perform in court as well as in the laboratory, a task that required very different skills. The adversarial British court system, in which witnesses were subject to inflexible yes/no questioning by barristers, proved to be a more challenging situation for expert witnesses than the civil law courts across Continental Europe, in which scientific experts could express the nuances of scientific evidence more clearly.[58] Concern over the less stringent interpretations provided by government scientists in disputed court cases under the Sale of Food and Drugs Act was a continual theme of the *Analyst*'s editorials.[59] The SPA urged Somerset House officials to meet with it, in order to negotiate common standards and approaches that would overcome the lack of consensus.[60] The creation of standards was an important feature of Victorian science and a means by which the state was able to monitor, protect, and control the population. During the nineteenth century, natural products such as food became increasingly subject to investigation and classification by scientists across Europe, while the increasing use of statistics helped experts and governments regulate food.[61] Codification by way of standards is as much political as it is scientific. As countries, institutions, and companies use standards for a variety of different reasons, such as the control of trade, the creation of standards or norms is subject to lengthy negotiation and compromise among many actors.[62]

Hamlin has observed the difficulties surrounding public analysts' struggle to establish standards for assessing water quality and purity, pointing out that in a profession "in which consensus on technical matters had profound legal implications, matters of standard methodology were of great import."[63] Contrary to the public analysts' criticism, the chemists at Somerset House were not against standards per se, and had in fact tried hard to establish standards for tobacco, beer, and milk.[64]

However, according to the laboratory's records, the government chemists recognized that the natural variations in perishable products made the setting of stringent and prescriptive standards often misleading and erroneous as a means of establishing adulteration.[65]

Many of the disputes between the government chemists and local authority chemists revolved around the fat content and watering down of milk. From the 1860s, the milk trade in the large industrializing cities of Britain had led to greatly lengthened supply chains, which increased the opportunities for middlemen to dilute milk with water, or skim off the cream, to increase profits. Such widespread practices encouraged public analysts to develop faster and more cost-effective tests. Wanklyn and other public analysts, most notably Kent's public analyst, Matthew Adams, developed their own tests. These tests, however, had large margins of error, which varied significantly depending on their use in different laboratories. The inconsistency of results and lack of consensus among analysts was remarked upon by fellow analyst Otto Hehner, who claimed that variations were due, in part, to differences in laboratory protocols, such as heating temperatures and times.[66] Hehner compared the milk-testing techniques of Voelcker and Wanklyn to those of government chemist James Bell and German-born Paul Vieth, chief chemist of the Aylesbury Dairy Company. Hehner observed that "each analyst, working in his own way, got results correct for his own way of working, but not comparable with those of any other analyst. We all found fat, and solids-non-fat; but we all attached different meanings to these terms, without exactly knowing it."[67] It was a situation that had been going on for years and was far from being resolved, according to Hehner, who noted that "whenever during the last ten years a number of analysts met, be it in Court, or at discussions of this subject, they differed and differed violently. They looked upon themselves as infallible; upon others as erring, or worse."[68]

The use of different analytical techniques to measure a product such as milk, which itself was subject to natural variations, was bound to produce inconsistent results, according to the government chemist James Bell.[69] Bell and his fellow chemists in Somerset House used a testing technique that involved macerating evaporated milk, a process that was considered too expensive and time-consuming by public analysts. However, the maceration method worked well for milk that had already started to decompose, a problem often faced by the government chemists of Somerset House, who, as final adjudicators, received samples later in the testing regime than the public analysts. Variations in results caused by different types of testing were compounded by different formulae used to compare the different measurements taken. The use of inconsistent equations and different experimental techniques, performed by different analysts working for different clients with different objectives is a good example of Andrew Pickering's "mangle of practice."[70] Indeed, Pickering's theory that changing interactions between experimental techniques, instruments, scientific theories, and humans are mangled together in different cultural contexts forms a thread running throughout the present book from the manufacture of organic dyes, to the creation of the tests to detect and assess them, to the rules and regulations surrounding food quality, and to the food industry itself.

Differences in opinion between the public analysts and the revenue chemists related to differences in standards and interpretation amidst an atmosphere of hostility and of lack of mutual respect and transparency.[71] The differences in analytical methodology and interpretation derived from the different interests and objectives of each chemist. The public analysts presented themselves as practicing chemists serving the health of the nation by rooting out fraudulent traders committing adulteration. That they sought universal standards in order to secure convictions in court is clearly suggested in Hehner's protestation that "it will be a scandal if we, as scientific men seeking the truth, and nothing but the truth, cannot agree upon a common method of measurement."[72] Historians such as Burnett have contended that the focus of the government chemists was very different, since food regulations were enforced by the Customs and Excise Department for tax reasons rather than health and safety considerations.[73] The government chemists, meanwhile, had performed extensive studies and experiments on the natural variability of milk and other food products and concluded that the establishment of stringent compositional standards on products that varied seasonally and regionally might "inflict a serious hardship on many honest vendors."[74] Recognizing the wide range of "normal" in commodities based on natural products, such as milk, the civil servants of Somerset House adopted a broader, more flexible approach to the establishment and application of standards, which contrasted with the exacting, prescriptive, and rigorously scientific approach advocated by the public analysts.[75] Steere-Williams claims that Bell and the government chemists also "avoided universal limits in order to expand their research program." The government chemists saw themselves as researchers, interested in learning more about the different qualities and variations of food, rather than regulators, ensuring that food items met fixed and pre-determined criteria.[76]

In their criticisms of one another, the two groups of chemists were positioning themselves differently. The public analysts were establishing themselves as professional consultants using chemistry to protect the public from adulteration or poisoning, or to defend companies from accusations of adulteration. While Somerset House's deputy principal, Richard Bannister, recognized the importance of antiadulteration legislation, more important to him was the ability to feed the population of Britain. To Bannister, free trade, the opening up of foreign markets, and improvements in science, technology, food preservation techniques, and food transport had allowed the provision of cheaper food for the masses, as well as more choice and higher and more uniform quality. Bannister accused public analysts of being too prescriptive in their efforts to define and regulate "natural" food products using standard scientific formulae. Studies by government chemists showed considerable variation between, for example, different tobacco leaves, making the ability to detect intentional adulteration not a straightforward exercise.[77] Analysts, however, desired uniform standards as a means of increasing both their authority and their ability to argue cases in court. Continuing disputes over methodology and the degree of natural variability in commodities made the search for standards highly contested.

The two groups of chemists accused one another of outdated and inaccurate techniques. In a public discussion in 1894 on proposed amendments to food legislation, Charles E. Cassal, public analyst for High Wycombe, Kensington, Lincoln, and later Westminster, declared that the "officials at Somerset House" should be relieved "from all connection with the Acts," with the question of reference being placed in the "hands of a new authority thoroughly acquainted with the most recent scientific facts and methods, and in a position to obtain the latest information." According to Cassal, "the Somerset House official laboured under the impression that, as a Government official, he was in every sense a very superior person." But Cassal doubted that the government chemists would ever improve, "bound down as they were, and would be, by the benumbing shackles of red tape, and the deadweight peculiar to an old-established and somnolent Government Department."[78]

The public analysts' constant criticism of the government chemists, as well as others involved in the food monitoring and production industry, tells us much about the analysts themselves and their struggle to establish their status and authority among fellow professionals. Looking back in 1932 on the early decades of the public analysts, former SPA president Bernard Dyer observed that the public analysts' sometimes intemperate language and bellicose behavior arose partly because of their passion for reform and noted that "it must be remembered that we were all going through troublous times."[79]

The analytical chemists who became state-sponsored arbiters of food reliability across Europe faced considerable challenges, as demonstrated in Britain. Looking behind their veneer of crusading zeal and projection of scientific prowess and precision, it is possible to glimpse a group of men struggling to forge careers and reputations as a new breed of chemist. The public analysts were a small group of chemists, some self-taught, others who had attended the new technical colleges in Britain before traveling to Germany to extend their knowledge. Many of them knew each other and had been taught by, or worked under, the same people. Unlike the government chemists, they were unable to rely on a life-long state salary and found themselves touting their skills to different companies, local authorities, and members of the public. The public analysts sought stricter standards and uniform methods to strengthen their authority and credibility and presented their role as that of protectors of the safety and morality of the nation's food supply. The government chemists, by contrast, adopted a more pragmatic approach to food production, recognizing the vagaries of natural and processed products, and with their primary aims being in line with the government's, those of respecting the free market, maximizing tax revenues, and enabling the availability of cheap food. The government chemists also viewed their scientific research as a means of achieving greater understanding of food composition rather than merely an instrument to prosecute adulterators.

Synthetic dyes were entering the food supply system during a period when public analysts were clearly fighting to secure their place as creditable and trustworthy analytical chemists, as well as seeking to persuade the public that chemistry was the foremost sanitary and socially useful science. The new dyes were among a range of chemical additives that quickly became commodified and valuable ingredients in food, within an industrial, commercial, and economic landscape of a nation wrestling with conflicting ideas surrounding the benefits and drawbacks of a free market versus state intervention.[80] At the same time that public analysts were being appointed to monitor the food supplied to the public, commentators such as the natural philosopher and political theorist Herbert Spencer were warning against the "development of the regulative apparatus" imposing rules and regulations on the "spontaneous" and "efficient" system of demand and supply that "allows great cities to be fed."[81] While the dairy industry lobbied against the coloring of margarine, Bannister, one of the government's chief chemists, argued that, as long as it was not passed off fraudulently as butter, colored margarine should not be discriminated against but rather welcomed as an important addition to the cheap food supply.[82] Unlike some of the toxic mineral–based dyes previously used to color food, there was no evidence, and certainly no consensus, about the risks or benefits of using the new chemical dyes. Indeed, as was demonstrated in the previous chapter, the ability of any chemist to detect the new dyes in food, let alone establish their safety, was almost nonexistent.

British analysts found it hard to assert their authority and prove their credibility as professional experts able to police the nation's food supply as they struggled to establish food standards and agreed-upon methods for proving adulteration for long-established food practices. Could it be that a lack of status, authority, consensus, and technical expertise as well as a desire to endorse the reputation of chemistry, particularly the type of organic chemistry emanating from Germany at that time, contributed to the reluctance of many British public analysts to pronounce judgment on the use of chemical dyes in food?

6

How British food chemists responded to the use of coal tar dyes

While the US government produced a list of recommended food dyes in 1907, and many European countries had prohibited certain coal tar dyes for food use by 1900, Britain waited until 1925 to ban a few specified coal tar dyes. A list of permitted coal tar dyes was not introduced into British law until 1957, fifty years after similar US legislation. So why was Britain one of the last Western countries to introduce legislation on these new chemical substances, despite being one of the first nations to introduce food legislation and to appoint public analysts to assess food?

Debates among public analysts and evidence presented to parliamentary committees suggest that most analysts in Britain were ambivalent about the use of synthetic dyes in food prior to 1890. They balked only if the dyes were deliberately employed by food manufacturers and retailers to deceive the public or were proven to be toxic in the small quantities used. While concern increased between 1890 and 1912, public analysts were divided as to whether regulation was needed to address the situation. Many still contended that the dangers were probably small, and certainly less than the dangers of many mineral dyes previously used as colorings. Analysts argued that any drawbacks associated with the use of coal tar dyes needed to be weighed against the benefits of consumer choice and the practicalities of production. As late as 1911, the entry on "Adulteration" in the *Encyclopaedia Britannica* argued that "the employment of powerful aniline dyes is *an advance* as compared with the use of the vicious and often highly poisonous mineral colours" (emphasis added) in the mid-nineteenth century.

There is little evidence that many British analysts were investigating the use of synthetic dyes in food or drink, let alone devising methods and experiments for their detection. However, there is plenty of evidence that they were reading in the British specialist press about efforts overseas to detect the presence and assess the toxicity of the dyes. Moreover, by the 1880s, the general press in Britain began to report on "poisonous" dyes in clothes and food.

Sir Charles A. Cameron, president of the Royal College of Surgeons and Public Analyst for the City of Dublin, was representative of how British analysts viewed the use of synthetic dyes in food in the mid-1880s. Describing the use of coal tar dyes in confectionery, he noted that:

> as the coal-tar dyes are liable to contain traces of lead, mercury and arsenic, and as their use for the purpose of colouring confectionery has been prohibited by the authorities in Paris, it would appear to be the safer plan not to place these dyes in any substance intended for human food. The quantity used, however, is so small, that no serious consequences are likely to arise from eating confectionery coloured with these substances.[1]

This highlights the prevailing opinion of British analysts at that time: that the toxicity of the synthetic dyes lay with poisonous contaminants such as arsenic and that their use was in such small amounts as to present no danger to the consuming public. However, by 1891, Cameron, like some of his peers, began to qualify his views. At a meeting at the Royal Academy of Medicine in Ireland, he stated:

> In cases of poisoning attributed to aniline colours, it was generally supposed that the toxic effect was caused by the arsenic which is often used in preparing coal tar colours. It is, however, not unlikely that certain of the aniline dyes may be somewhat toxic, though quite free from arsenic. I have met with some cases which rather favour this supposition. Last August, a gentleman residing in Lesson Park, Dublin, sent me some large comfits, the use of which he believed had made his children ill. Three young

children, after partaking of, as alleged, a moderate quantity of the comfits, became very ill, with severe vomiting and purging, and considerable prostration. The comfits were of a very deep red colour, and the pigment was laid on them in greater quantity than, I think, I have ever noticed before. No arsenic was detected, and the colouring matter proved to be the aniline dye, termed fuchsine. . . . About the same time several children were treated in Mercer's Hospital for symptoms which appeared to be due to an irritant poison. In this case the suspected poison was a sweetmeat in which I found a thick coating of aniline dye. About a fortnight ago several children became very ill after eating some bright red confections; they suffered from vomiting and purging. In this case the confections were coloured with fuchsine.[2]

However, rather than advocating more experiments either to detect the presence of such dyes in food or to assess the toxicity of aniline dyes, Cameron suggested that analysts keep a watchful eye over any future cases where poisoning by aniline dyes might be suspected, noting that "the above cases cannot be regarded as conclusively proving the toxic nature of aniline colours in food, but it is worth while recording them as further cases of the same kind, if they occur, might convert doubt into certainty."[3]

As already demonstrated, the state of knowledge about, and ability to test and assess, the dyes being used in food at this time was far from certain. British analysts also were struggling to assert themselves as professional experts whilst operating within the laissez-faire, free-trade political economy of Victorian Britain.[4] Public analysts were appointed by local authorities to represent the interest of the consumer, but many were also paid to represent food producers and retailers. Public analysts, consumers, social reformers, free and fair trade advocates, manufacturers, and retailers all had a stake in forming economic and cultural knowledge, and all invoked public health and consumer rights arguments to further their aims.[5]

By the late 1890s only a few campaigning analysts, most notably Charles Cassal and Otto Hehner, begin to call for bans or limits on chemical additives, including colorings and preservatives. More strident than most of his peers about this issue, Cassal resigned from the Society of Public Analysts following the International Congress of Hygiene and Demography in Vienna in 1889, and became an active member of the International Commission on Adulteration. In 1899, together with his mentor, the hygienist William Henry Corfield, professor of hygiene and public health at University College London, Cassal set up *The British Food Journal and Analytical Review*, which he edited until 1914, to campaign for greater food purity and regulations.[6] Cassal's contemporary, Otto Hehner, was public analyst for Nottinghamshire, West Sussex, the Isle of Wight, and Derby and one of several German-born and/or trained chemists working in Britain. Hehner rose through the ranks to become president of the SPA and vice president of the Institute of Chemistry. Trained originally under the analytical chemists Carl Remegius Fresenius and Carl Neubauer at the Wiesbaden Agricultural Institution in Germany, Hehner moved to Glasgow's Andersonian College, where he worked with Gustav Bischof and William Ramsay, before eventually becoming a partner in a consulting practice set up by Hassall. He became one of the country's leading food analysts while keeping abreast of developments throughout the chemical sector, including the industrial production of organic dyestuffs. Hehner's views and comments on the use of chemical dyes varied over his working life. However, by the end of his career, Hehner became increasingly vociferous in his criticism of the use of chemicals in food.[7]

The use of coal tar dyes in margarine was one of the most contentious areas of food coloring in nineteenth-century Britain, one in which public health and economic arguments became intertwined. In many ways, margarine encapsulated the issues surrounding food production during this period. While promoted as an economic and hygienic new food product offering more choice to consumers, particularly those with less income, its novelty presented uncertainty and a threat to established interests and norms.

Oleomargarine was a novel food invented by the French chemist Hippolyte Mège-Mouriès in 1869, in response to a request by Napoleon III to find a cheap substitute for butter for the armed forces and the poor.[8] Originally based on beef fat, oleomargarine was naturally white in color, but manufacturers and retailers increasingly dyed it yellow to make it appear more like butter and to be more acceptable to consumers. Production of margarine, a term used to describe oleomargarine and other fat substitutes for butter, increased dramatically during the later decades of the nineteenth century.

According to the 1901 governmental report on Food Preservatives & Colouring Matters, annatto and other vegetable-based colors such as turmeric and saffron, widely used to color butter and milk, were "being superseded by coal-tar yellows, the action of which upon the human system is not fully known." The report noted that "butter from Holland,

Australia and the United States is very frequently coloured with coal-tar yellows" while "a large number of margarines are also coloured." The dyes most commonly used were dimethyl-amido-azo-benzene, known commercially as "butter-yellow," and tropaeolins, sulphonated-azo derivatives from coal tar. Butter-yellow was generally supplied to the trade already dissolved in oil, such as cottonseed, rape, linseed, or sesame oil.[9]

The same report pointed out that the highest usage of coal tar colorings in food generally occurred in margarine: the dyes were in 100 of 133 of samples analyzed by the Government Laboratory in Somerset House. Much of the concern over the coloring of margarine was raised by the powerful dairy industries. But even while butter producers wanted to ban the coloring of margarine, they themselves used artificial colorings such as annatto, aniline, and naphthol dyes to ensure consistent coloring in their own product, as the color of butter tended to vary during the year. As is often seen in food production and trade, public health arguments were invoked primarily to settle commercial disputes.

Within a few years of margarine's introduction, governments from Germany to America were legislating against the new product, with most objecting to its being colored yellow. This legislation was based on protecting the public not from harm but from deception. The new laws also helped to protect indigenous dairy industries from competition, both domestic and foreign. In Britain, the government and analysts took a different approach from the US, Canada, Australia, and other European countries in their response to margarine being colored yellow. Although British analysts were concerned that the use of dyes in margarine might fool customers into thinking they were buying butter, most chemists did not believe that yellow coloring should be banned outright, nor did they differentiate between the use of vegetable dyes, such as annatto, or coal tar dyes.[10]

In a Bill to Regulate the Importation, Manufacture and Sale of Butter Substitutes (Margarine Act 1887), the British Government did not restrict the use of coloring in either butter or butter substitutes, merely requiring that all margarine be labeled as such. First offenders against this law were to be fined £20, with second offenders facing up to one month's imprisonment or a £50 fine and subsequent offences resulting in up to six months imprisonment. The Margarine Act determined "that the word 'butter' shall mean the substance usually known as butter, made exclusively from milk or cream, or both, with common salt, and with or without additional colouring matter . . . [while] . . . the word 'oleomargarine' shall mean all substances, whether compounds or otherwise, prepared in imitation of butter, and whether mixed with butter or not, and no such substance shall be lawfully sold, except under the name of oleomargarine."[11]

The highly punitive measures included in the legislation demonstrate the power of the dairy industry in Britain at the time. Dairy producers effectively obtained trade restrictions on margarine by invoking public health and economic fraud arguments. However, the butter producers were against any measures to ban the use of artificial coloring itself and turned instead to arguments advocating consumer choice to defend the use of coloring in both margarine and butter. This was a very different approach from that adopted by the US butter producers, reflecting the laissez-faire commercial and political culture of Britain compared with the protectionist culture of nineteenth-century America. Robert Gibson, salesmaster of the public creamery markets in Limerick, explained how food tastes varied geographically, leading to many butter producers arguing that "it is absolutely necessary to have colouring to suit the tastes of different districts." Gibson observed that the people of Liverpool, Manchester, and Oldham all liked their butter slightly different shades and "when you come down here into the south they will not take pale butter from you at all." As a result, producers and wholesalers "that have a scattered trade, trading north, south, east and west say that the colouring is absolutely necessary to meet your markets."[12] As we shall see later, this tension between local consumer tastes and regional variations in food production and the desire for universal scientifically measurable standards was not unique to Britain.

Many grocers also argued that excessive legislation on food coloring would deny consumers a choice. The Grocers' Federation was "opposed to an interference with the colouring of margarine, and to the prohibition of the mixing of margarine and butter, on the ground that neither course would necessarily prevent fraud, whilst the prohibition of mixtures would increase the price of butter, and seriously affect the trade in margarine, thus depriving the poorer classes of an article which is wholesome and nutritious."[13] In 1892, *The Public Analytical Journal and Sanitary Review*, as part of its campaign for a Margarine Act to prevent margarine being sold butter- color, criticized grocers for not supporting such a bill:

> Margarine is coloured its butter tint for purposes of deception. The substance should be sold in its natural colour, as in Denmark, and then the public who want margarine would know that they get it, whilst those who want butter would be protected from having margarine palmed upon them as butter. To

our thinking, the grocers opposing the amended Margarine Act are signally ill-advised, and are injuring their reputations for honesty.[14]

The article condemned the grocers' trade journals, claiming that they were full of advertisements for margarine and butter mixtures and "of adulterated articles of every kind," and proclaimed that grocers should stand up against the practice:

> it is not in their interest to encourage adulteration in any form; and it is high time the grocers, as a body, gave some serious study to the question. They would quickly see the mean, sordid motives of their so-called trade journals in egging the various federations on to the defence of dishonesty. Their trade journals are reaping colossal fortunes out of the advertisements of manufacturers of adulterated articles, and as such manufacturers pay the piper, the trade journals find it to their interest to let their paymasters call the tune. This, in a nutshell is the whole and sole reason of the resistance to an amended Margarine Act, and, we repeat, is utterly unworthy of the grocers of this country.[15]

Editorials in the trade, professional, and consumer press often accused their competitors of unprofessional practices, particularly relating to the sale of advertising. The fact that many of these magazines and indeed professionals themselves, including public analysts, relied on food manufacturers for part of their income was a situation often commented upon. The SPA regularly debated the issue of whether or not public analysts should be allowed to accept financial rewards for endorsing food and drink products. While many analysts saw margarine and other industrialized food products and ingredients as efficient, effective, sanitary, and scientific answers to the vexed issue of feeding a rapidly growing and urbanizing population, the fact that analysts were paid by producers to assess and endorse their products jeopardized their ability to stand for disinterested true chemical knowledge and discredited their claims to be impartial arbiters of the public food supply.

Articles in the sanitary press rebuked public analysts and the SPA for not campaigning against the sale of colored margarine, and also for making money by promoting margarine as healthy and clean:

> Unfortunately some eminent scientists have been found to puff any substances [and] sell their names and honour, and to write

advertising puffs for any rubbish for "a consideration." They belong to the class of scientists who are in the pay of, and bound body and soul to, various trade associations, and who can be relied upon at any time for "a consideration" to use their unquestioned skill to shield their employers from punishment, however great be the adulteration fraud that honest public analysts strive to put down.[16]

Public analysts were caught in the middle of debates between different interest groups all invoking public health and economic fraud arguments to further their interests. Many public analysts also argued that banning the use of artificial coloring in food would reduce consumer choice. August Dupré, public analyst for Westminster, claimed that the public had become accustomed to their food being colored and in many instances preferred it: "the public do not like discoloured peas. In a great measure it is due to the fact that people do not like on their table a preserved food. Just like the poor, who do not want to buy margarine; they want to buy butter, or they do not want their neighbours to know they are buying margarine."[17] Most analysts were primarily concerned about consumers being duped into buying margarine thinking it was butter, or buying butter with low fat content. If the public preferred to eat yellow margarine rather than white margarine, that was not a problem, analysts argued, so long as the product was sold and labeled as margarine. Otto Hehner, however, complained that food producers were using the arguments of consumer choice as an excuse to artificially color their food for reasons of convenience and cost. Public analysts, he declared, should "disregard the popular wish and raise the standard of purity of food," arguing that "the popular wish had been used by sophisticators for many years past as an excuse for every abomination" and that "no kind of adulteration was ever carried out without its being alleged to be in obedience to the public wish."[18]

The historian Frank Trentmann has shown how public health campaigners and producers and retailers in the nineteenth century all invoked "consumers" and the "consumer movement" to serve their own purposes.[19] By shaping themselves as protectors of the food supply, public analysts claimed that they, rather than the food industry, were the most appropriate experts to determine the rules and regulations over what constituted legitimate and illicit food intervention and how food additives should be identified and categorized. However, the nineteenth-century debates on adulteration and food chemistry show that knowledge was not the critical factor in establishing expertise. The lack of consensus as to what constituted adulteration and the difficulty of knowing and identifying synthetic dyes and their toxicity meant that experts fell back on their authority and moral and political views. While much historical work has been done to examine how scientists asserted their authority to construct facts and consensus in fields such as metrology and electricity, analytical chemistry arguably offers a more complex and interesting account of experimenters' regress and the social negotiation of "scientific truths."[20]

The difficulties chemists faced in determining "facts" and "truths" about the new dyes in the laboratory have already been shown in chapter 4. The negotiation surrounding, and understanding of, these enigmatic substances became even more complicated once they left the laboratory and entered the commercial marketplace. Part of the problem lay with the public analysts, whose technical and professional insecurities, combined with their having to work for different paymasters, led to a lack of consistency in their approach to the coal tar dyes. As will be shown, even Hehner and Cassal, two of the most vociferous critics of artificially coloring food with coal tar dyes, often contradicted each other and themselves as they adjusted their rhetoric and opinions to suit their audience and situation.

Coloring Sugar

The degree to which public analysts constructed their versions of the truth, depending on where they were operating and on whose behalf they were acting, can be seen by examining their response to the coloring of sugar. Sugar began the nineteenth century as a luxury, an exotic import from Britain's Caribbean colonies, and ended as a ubiquitous and affordable food commodity manufactured from both home-grown beet and colonial sugar cane. Prior to the nineteenth century, highly processed white cane sugar had been the most desired form of sugar. However, as beet sugar began to be increasingly manufactured in Europe, with its chemically similar, processed, white crystalline grains allowing it to be passed off as white cane sugar, consumers began to favor the distinctive golden-brown Demerara sugar. "Demerara sugar" is a term used to describe both a place of origin in the Guianas off the northern coast of South America and a method of production. Demerara sugar was one of several types of brown sugar produced in the Caribbean, distinguishable from the more refined white Caribbean sugars in that they retained more of the brown molasses syrup extracted from the white sugars. Like bread, the desire for white or brown sugar has fluctuated throughout its history, with scientific claims used by all interested parties to their advantage. As the popularity of "brown sugar" increased, British beet sugar refiners began to introduce artificial yellow dyes into white sugar, a practice not unusual in an industry where blue dyes were already used to increase the "whiteness" of white sugar. The beet sugar manufacturers also launched a campaign to smear brown sugars from the Caribbean, using microscopes and photography to show the presence of microbes in the less processed sugars.[21]

The coloring of sugar soon became a vexatious subject in British courts with the same professional chemists employed as public analysts, consultants to the sugar industry, and expert witnesses appearing for both prosecution and defense. By comparing private conversations with analysts' statements in court and their experiments in the laboratory, one can clearly see how public analysts attempted to construct both "knowledge" and "expertise." The evidence also demonstrates that the uncertainty surrounding adulteration provided opportunities for public analysts to earn money as consultants on both sides of any adulteration prosecution.

The extent to which the public analysts had to devise strategies to ensure that uncertainties did not undermine their credibility as experts or the status of chemistry is clearly shown in a meeting of the SPA in London in 1890, reported later in the *Analyst*.[22] The analysts' inability to accurately test for the dyes, as described earlier, made courts, in particular, a worrying place for them. During the discussion, Cassal advised his fellow analysts that, in court, it was less important to identify *which* aniline dye was present than to determine *whether* any dye had been used at all. In fact, he argued, it was undesirable and unnecessary to make too specific statements as to the precise nature of the dyes used. For if an analyst specified an exact composition or ingredient it became easy for the manufacturer or retailer to swear in court that a particular compound or chemical had not been used, when it was not exactly identical to the one that the public analyst claimed to have been the adulterant. August Dupré agreed that "it would be a most dangerous thing for the public analyst to bind himself down to a particular composition." The analysts argued that it was the duty of the analyst to prove adulteration, rather than detail the exact chemical composition of the adulterant. This cautious approach stemmed from the chemists' inability to make verifiable claims about the dyes. Cassal stated that it was always inadvisable to go into technical details before a magistrate because of the court officials' lack of scientific training, noting that most were "quite incapable of judging of the merits of the most elementary scientific questions."[23] These comments show that chemists were aware of the differing standards of proof required in the courtroom and the laboratory, standards based not so much on the credibility of the facts themselves but on the levels of expertise of the legal profession. In a clear example of public analysts creating and policing their professional boundaries, Cassal justified the analysts' failure to detect individual dyes by suggesting that such complexity was beyond the capacity of the court officials' comprehension or the needs of the process. At the same time, analysts knew that their powers were limited before these pervasive but chemically elusive substances, and that there was a high risk that their public statements could be easily contradicted in the absence of experimental proof. This exchange makes explicit the maneuver of falling back on socially constituted expertise when experiments fail and demonstrates how worried the analysts were about their performance in court and why different rhetoric was needed.

Disagreeing with Cassal, however, Otto Hehner told his fellow public analysts that they had a responsibility to check the use of poisonous substances in food and, therefore, should attempt to identify individual dyes used. Hehner recognized that the debate over the new chemicals being used in food would benefit the public analysts' efforts to carve out a specific kind of role distinguishing them from other chemists. Their ability to do this rested on setting themselves up as disinterested public figures with a broad-ranging knowledge comprising chemistry and physiology. Hehner argued that several aniline colors, including dinitrocresol (also known as Victoria yellow, among other names) and dinitronaphthol (Martius yellow), were "of a decidedly poisonous nature" and that it was the duty and expectation of public analysts to "know the exact nature of the colouring matters employed." Clearly positioning the public analysts as the neutral arbiters and experts in what constituted healthy ingredients, Hehner stated: "just as a public analyst might not need to know all details of technical and manufactory processes, so manufacturing chemists might possibly be ignorant of the physiological action of the colours which they added to food materials."[24] Dupré also laid claim to public analysts' sole jurisdiction over the validity of food additives, arguing that the Society of Public Analysts "as a scientific body should not accept the assertion of anyone, whoever he might be—let sugar manufacturers assert ever so much that the colouring matter they added was innocent—he should not pay the slightest regard to it unless they were prepared to state the nature of the colouring matter added, and then let him judge whether or not it was injurious to health."[25]

Hehner's comments related to concerns surrounding the possible toxicity of the coal tar dyes, rather than the issue of deception. Although Cassal agreed that many of the coal tar dyes might well be poisonous in large quantities, he argued that the main concern of most analysts was the use of the colorings to disguise a food product or to deceive the consumer. Dupré claimed that he often had artificially colored sugars submitted to him, noting, however, that he would be doubtful about stating such sugars to be adulterated unless they were being passed off as something they were not, such as beet sugar sold as Demerara. Dupré questioned how far analysts could obtain convictions under the Sale of Food and Drugs Act in cases of mere coloring if the coloring matter was harmless, or could not be proven to be harmful.[26] These comments demonstrate the problematical function of additives such as dyes, and the difficulties attached to assimilating them into the category of "adulterants."

These differing views held by public analysts on the role of artificial coloring and the issue of commercial deception were frequently apparent in the courts. Indeed, the public analysts' private debate described became public in a high-profile case held in Birmingham's Victoria Courts that highlighted the complexity of the issues surrounding the analysis and legitimacy of food ingredients. Here the question at stake was the nature of the product being colored, as much as the nature of the coloring matter used. When Thomas Davis, a grocer, was charged with selling dyed sugar beet crystals as Demerara sugar, both the prosecution and the defense employed three leading public analysts apiece as expert witnesses, putting the analysts center stage and at odds with each other.[27] As well as being paid as public analysts, each of these individuals had his own private consultancy business, and all were accustomed to acting for whoever hired them, whether public authorities, consumers, or food retailers and producers.

Appearing for the prosecution, Alfred Hill, public analyst and medical officer of health for Birmingham and a former SPA president, argued that because the sugar was dyed with aniline, it was not therefore likely to be Demerara. Although Hill was unable to guarantee that that the sample in question was *not* Demerara, he informed the court that he had analyzed 121 specimens of what he believed to be Demerara sugar during the previous nine years and had only found one with traces of aniline dye. Yet as already shown, analysts were not always able to detect the presence of aniline or azo dyes in food. This was a detail they were unlikely to dwell on in court. Also appearing for the prosecution, Cassal took a moral stance claiming that natural Demerara sugar ought

to be free "from any foreign substance," and warned the court that dyeing allowed inferior sugar to be sold as Demerara. When the defendant's
lawyer suggested that aniline dye might be added to Demerara sugar
to make "the sugar of a uniform colour and more pleasing to the eye,"
Cassal replied that this "was not the object he attached to the process
of dyeing." Alfred Henry Allen, public analyst for the West Riding of
Yorkshire, gave similar evidence. Trade experts were also called by the
prosecution; they stated that Demerara sugar was not a generic name
for any sugar from the Caribbean, but a specific term understood to
mean "pure cane-sugar, undyed, and coming from Demerara." They argued that dyeing inferior sugar allowed retailers to make more money.
While the chemists were unable to determine chemically whether the
sugar was Demerara or beet sugar, they turned to moral arguments of
"purity" and "deception" to make their case.[28]

However, the grocer's defense team, including three chemists, argued
that it was not an offense to dye real Demerara sugar, as the dye did
not alter the "nature, substance or quality of the sugar," and pointed
out that there was no proof that the sugar was not Demerara. It was an
offense to color or stain a food only when this practice made it injurious to health, the defense argued. Hehner, testifying for the defense,
claimed that dyeing West Indian sugar was a common practice and that
the aniline dye used was in his opinion harmless to health. A dye was
used in the preparation of nearly every article of food, Hehner added;
in the case of sugar the whole object of dyeing was to make the sugar of
an even color and pleasing to the eye and in accordance with the public
taste. Interestingly, these trial arguments give no indication of Hehner's
antipathy for artificial colorings. In fact, they contradict his comments
in private discussions among other analysts and in the sanitary press,
where he spoke out critically about the dangers and deception of using
preservatives and colorings in food. By the end of his career, Hehner was
thoroughly disenchanted with the industrialization and manipulation
of food with the addition of chemical preservatives and colorings. In
1923, he would even be removed from a Ministry of Health departmental committee on preservatives, after writing a letter to *The Times* ex-
pressing his forthright opinions on this very issue before the committee
met.[29] He would become, along with Cassal, one of the most ardent and
outspoken campaigners against the use of chemical colorings. Hehner's
comments in court thus demonstrate the capacity of consulting chemists
to adapt their professional opinions to meet the criteria of their clients,
whatever their personal views.

Comments by his fellow defense expert Benjamin Newlands, a consulting chemist who worked closely with sugar refiners, also raise the question of whether expertise can ever be politically or economically neutral. Newlands confirmed Hehner's observations that aniline dyes were frequently used to color sugar, both because of consumer preference and also because aniline was considered a safer dye than other substances used to color sugar. Newlands claimed that he had himself introduced the system of dyeing sugar with aniline dyes, following the sensation caused by a rumor of poisoning by the use of chloride of tin to darken sugar.[30] Newlands confirmed that he had manufactured sugar in London by "the Demerara process," in which white sugar is colored brown, but would not admit to dyeing white beet sugar. Like many consulting analysts at the time, Newlands had a close working relationship with the food industry. He and his brother John had their own analytical consulting practice and worked with sugar refineries and were co-authors of a handbook on sugar for planters and refiners.[31] Newlands's son-in-law, John Joseph Eastick, had set up a sugar analysis business with his two brothers in London, later becoming chemist at the sugar refinery of Abram Lyle in the early 1880s, where he developed the first formula for golden syrup.[32]

Bernard Dyer was the third analyst to give evidence for the defense, providing a full counterweight to the trio of high-profile public analyst experts called by the prosecution. To make the case for the retail trade, the defense called on Councillor Jarvis, vice president of the Grocers' Federation, to contradict the prosecution's "trade experts" by stating that, in the retail trade, Demerara sugar was understood to be any West Indian crystallized sugar cane and that it was understood that it was dyed to give it its "complexion." The magistrates dismissed the case, judging that while the sugar was dyed, it was of the same quality as Demerara, and that the dye was not injurious to health.

Golan and Hamlin claim that men of science primarily took work as expert witnesses to boost their income at a time when professional men struggled to earn a living from science.[33] While this is certainly the case, the involvement of British public analysts in the arbitration of food adulteration was much wider than just financial interest. Indeed, the evidence also corroborates Carol Jones's contention that chemists and other men of science appeared as scientific experts in courts, scientific advisers to government committees, and scientific commentators in the Victorian media in order to ensure that science and scientists played an active role in social and political decision-making.[34]

Parliament investigates

Increasing public concern surrounding the use of chemical additives in food, including synthetic colorings, prompted Parliament to establish a Parliamentary Select Committee on Food Products Adulteration specifically to "inquire into the use of preservatives and colouring matters" and determine whether their use was "injurious." The committee's 1896 report confirmed that chemical dyes were widely used in food products and suggested that some limit on their use might be adopted. However, the subsequent 1899 Sale of Food and Drugs Act legalized preservatives and colorings used in the preparation of food without imposing any limits on their use. This had the conflicting effect of both further legitimizing the use of chemical dyes in food while at the same time raising consumer concerns as to their widespread use in food.[35]

The use of select committees, which comprised a group of cross-party parliamentarians questioning a cross-section of representatives from industry, the public, and health and scientific advisers on an issue, such as the use of food additives, was one way in which the government sought to mediate and obtain a balance between conflicting interest groups. A new committee listened to twenty-six days' worth of testimony presented by seventy-eight witnesses, including several prominent public analysts and food manufacturers.[36] The committee's formation reflected an increasing tendency for parliamentarians in the second half of the nineteenth century to seek "scientific advice" on matters of public health, a role the new breed of professional men of science were eager to undertake in order to boost both their status and reputation and that of science.[37] In their study of food regulation in Britain between 1875 and 1938, French and Phillips argue that it was public analysts who pressed for regulation on chemical additives, particularly preservatives, and who were most vociferous against their use in food. However, while there were a few outspoken critics of chemical food colorings among analysts, most prominently Cassal, Hehner, and Hill, their inability to detect coal tar dyes in food and assess their harmfulness led to a lack of consensus among analysts and hampered attempts to regulate their use.

Indeed, even the more ardent critics of coal tar food coloring toned down their protests before the select committee. Analysts seemed to be telling different stories in different settings, adapting to their multiple and often conflicting roles as public and political advisors, food custodians, scientific researchers, and commercial consultants. Hehner's testimony before the select committee, like his intervention in the earlier Birmingham court case, once again seemed to contradict the more strident views he had expressed in private among colleagues and in his comments, often contributed anonymously, in the sanitarian press of the period. Hehner told the committee that he had noticed an increasing use of aniline dyes in butter but was not concerned because the quantities used were so small. When asked about the use of Martius yellow, he observed that he had never come across it but admitted that the size of samples was so small its presence would be difficult to ascertain.[38] These comments contrast with his observations at the SPA, where he spoke of Martius yellow being "eminently poisonous."[39] For someone who campaigned vociferously, though often anonymously, in the press against the increased use of chemical ingredients in food, Hehner took a distinctly conciliatory approach at the parliamentary hearing, telling the committee that coloring matter in milk is "continually present—almost invariably in the London milk." While critical of the practice, he did not advocate banning it: "I think it is a deceptive practice, and undoubtedly was due originally to the milkman being anxious to hide the blueness of his milk. At the same time, it is such a universal practice now that the consumer would probably refuse the natural milk if he got it . . . I do not like the practice. At the same time it is not a matter which I think calls for urgent attention."[40] In a setting where prominent food manufacturers were also present, and where witnesses' testimony was being widely reported in the press, one wonders how much Hehner's public comments reflected his desire to continue working as a commercial consultant, rather than his personal point of view.

According to the Government Laboratory, coal tar colors were being increasingly used in food and drinks, especially in margarine (found in 100 of 133 samples tested by the laboratory), sauces and ketchups (found in 5 of 10 samples), cordials (12 of 24), and fruit syrups (12 of 23), but also widely used in sausages (72 of 226), butter (40 of 364), fruit jellies (9 of 28), potted meats (27 of 165), temperance drinks (56 of 769), and sugars (24 of 149). The government select committee noted that these figures probably significantly understated the actual use of coal tar dyes because of the difficulty of identifying the dyes in food and drink, given the minute quantities and varying strengths in which they were used. They were the harder to identify because the "characteristic reactions for the colouring agents are frequently interfered with or obscured by the organic matter of the food itself, especially after decomposition has commenced."[41]

The widespread use of the dyes, in tiny amounts that made them almost impossible to detect and that tended to produce a risk of cumulative poisoning rather than making individual products toxic, was a challenge for regulators and analysts. To ban all coloring of food would be viewed as restricting consumer choice. However, attempting to determine safe and acceptable thresholds for the multitude of different dyes was beyond the technical capability of analysts and threatened their credibility and authority. Problems that analysts and government chemists encountered in their attempts to establish agreed-upon standards for the composition of staple products such as milk were difficult enough. Trying to establish the impact of the enigmatic and largely undetectable coal tar dyes in food was a far more problematic matter, making analysts reluctant to hang their expertise and credibility on discrediting the dyes as a matter of public health.

Analysts, like the food industry and consumers, were divided as to how to respond to these novel and labile substances, created by chemists but manipulated outside of the laboratory in ways that made them hard to identify and even harder to manage. The difficulty of trying to determine whether the use of tiny amounts of certain ingredients could be detected, and whether they should be regarded as adulterants or not, was described by government chemist Richard Bannister, a director of Somerset House, at an earlier hearing of the House of Commons Select Committee on Food Products Adulteration in 1894.

Q. 2686. *If coloured or yellow crystals are sold as Demerara sugar, ought the sellers, in your opinion, to be prosecuted under the Sale of Food and Drugs Act for adulteration, or under the Merchandise Marks Act for giving a false trade description?*—It strikes me that it should be under the Merchandise Marks Act, so long as the colouring-matter is so small that it does not amount to adulteration.

Q. 2687. *You do not consider that the colouring process is an adulteration except beyond a certain point?*—I have examined a sample that has been coloured, and the amount of adulteration by colouring is so small, that it could not be considered adulteration.

Q. 2688. *Where do you think adulteration begins?*—It is very difficult to say where one begins and the other ends; it depends entirely upon the article itself.[42]

Analysts giving testimony to the 1899–1900 parliamentary committee agreed that detecting the tiny amounts of coal tar dye used in food products was difficult but also used the opportunity to urge government to provide more resources for research.[43] Walter William Fisher, the public analyst for Oxford, Berkshire, and Buckinghamshire and a chemical demonstrator at Oxford University, confirmed that the quantity of synthetic dyes "used is so small that it is as a rule impossible to absolutely identify any particular dye. It is say, one in 100,000, or 200,000 or 300,000 only." He informed members of the committee that dyes such as tropaeolin, as well as other nitro-compounds known to be toxic, were used in butter and eosin in jellies, but questioned whether such dyes used in such tiny quantities could be harmful. Fisher suggested that it would be helpful to draw up a schedule "as in Belgium" clarifying what was considered toxic, pointing out that Britain's "men of science" would appreciate funding for experiments on such substances for toxicity.[44]

Alexander Wynter Blyth, one of the country's preeminent public analysts, also told committee members that the use of such tiny quantities of coal tar dyes and the fact that analysts were usually presented with small samples of suspected food made it difficult for public analysts to identify colorings in food. However, Wynter Blyth expressed concern about the cumulative impact created by the extensive use of coal tar dyes in food and drink as a cheap replacement for dyes previously used, such as cochineal. While he agreed with his peers that the amount of dye used in any one item of food meant that it was unlikely to have an injurious effect, he questioned the cumulative impact of such colorings, particularly on children: "I cannot say that in any one substance that I have ever examined the quantity of aniline dye, even presuming it is poisonous, would be enough to injure health; but then when you consider that so many things are coloured in this way it is a question whether the collective amount that a child, say, might take in the day might not have some injurious effect." Wynter Blyth pointed out the work that had been done by both German and French chemists to determine the toxicity of some aniline dyes, proving several to be poisonous, but stated that it would not be possible at the current time to produce a list of injurious and harmless colors in Britain, "I do not think it possible, but I think it ought to be done." Other analysts also called for more government spending on chemical research to enable legislators to identify which toxic chemicals to ban.[45]

Hill recommended that Britain follow the example of France and Belgium in prohibiting certain colorings, invoking the analysts' public health responsibilities: "I think it is desirable to protect the public from having such drugs administered to them against their will, or against their knowledge." The resources being devoted in Britain to testing the toxicity of dyes in food were, however, minimal compared to Europe, where the practices of histology, physiology, and animal testing were more prevalent.

Cassal argued that "it should be the business of the legislature to ensure that the necessary information should be obtained at the expense and under the direction of the Government," claiming that the determination of whether particular dyes were toxic or not in the amounts used should not be put upon the "shoulders of individual public analysts" to fight in court.[46] According to Cassal, there was no definition of a poison or a drug under the 1875 Sale of Food and Drugs Act and that executing the act by proving a substance or item of food to be injurious was a costly and difficult business.[47] Several European countries had already banned certain coal tar dyes understood to be toxic, and the US would soon produce a list of permitted dyes, reducing the onus on analysts to prove toxicity in the law courts.

At the same time, issues of consumer choice and economic fraud also played a prominent role in the hearings, with analysts arguing that the British public had the right to consume what they wished, so long as they were not deceived. Even Cassal, normally an ardent campaigner against artificial dyes, accepted the use of color in butter, on account of "long custom" or "trade continuity."[48]

After listening to the testimony, the select committee concluded:

> In regard to the colouring matters of modern origin, while we are of opinion that articles of food are very much preferable in their natural colours, we are unable to deduce from the evidence received that any injurious results have been traced to their consumption. Undoubtedly some of the substances used to colour confectionery and sweetmeats are highly poisonous in themselves; but they are used in infinitesimal proportions, and before any individual had taken enough of colouring matter to injure him, his digestion would probably have been seriously disturbed by the substance which they were employed to adorn.[49]

The committee noted that the use of artificial coloring was long established in the butter and cheese trade and decided that it would not be in the interest of the consumer to interfere with the customs of these trades. The issues were, however, different for margarine, a new product with no long-established customs. While the committee recognized the

risk of margarine being passed off as butter, it did not consider the risk
to public health sufficient to prevent it from being artificially colored.

> In regard to margarine, we have to deal with a cheap and rela-
> tively inferior article coloured to resemble a more costly and
> superior article, and probably the only means of protecting the
> public from imposition would be to prohibit the introduction of
> any colouring into margarine which shall cause it to resemble
> butter. . . . But as the margarine may be assumed to be a per-
> fectly wholesome article of diet, it does not fall within the terms
> of our reference to make any recommendation upon a practice
> which is not attended with risk to the public health.

As far as milk was concerned, the committee did consider that action
should be taken, recommending banning use of coloring in milk "because of large quantities consumed and expectation among consumers
that it is a 'natural' product."

The response by the committee reveals a restrained negotiation
between freedom of trade, consumer choice, and the support of commercial practices versus consumer protection and greater transparency.
While defending the status quo, the committee also recognized that the
issue was by no means resolved and was not likely to go away, concluding that:

> the departmental machinery for controlling the preparation and
> conservation of food and drink in this country is not as com-
> plete as could be wished. The obvious fact has been referred to
> by several witnesses, that new methods of preserving and new
> preserving agents and colouring matters will continue to be in-
> troduced. We regard it as a matter of concern for the public
> health that the nature of such substances or processes should be
> critically examined and their effects upon the human economy,
> if possible, entertained.[50]

Britain's food legislation did not specify the use of coal tar dyes at
all, whether as permissible or prohibited. Its laws, enacted during the
second half of the nineteenth century, merely stated that food producers
and retailers should not harm or deceive the public without specify-
ing particular substances or additives that could or should not be used.
Legislation appointed public analysts to monitor the system with any

disputes being resolved in the law courts or by the government chemists at Somerset House. The courts, Parliament, and laboratories were all sites where food was monitored and policed in an expert system determined through compromises reached between various communities with different vested interests. In effect, the combined action of both parliamentarians and chemists, rather than controlling the use of synthetic chemicals in food, actually helped ensure that the use of chemicals in food production became further normalized and legitimized as industrial food processing increased, irreversibly changing our relationship with food.

7 French and German chemists seek to arbitrate the use of synthetic chemicals in food

While British public analysts were unwilling to pass conclusive judgment as to whether synthetic chemical dyes, or indeed any particular additive or substance, were harmful or acceptable as food ingredients, French and German chemists took a different approach. Unlike in Britain, chemistry had a high profile in political, educational, and industrial circles in both France and Germany. Chemists in these countries saw the introduction of chemical dyes in food as both an opportunity and a risk for chemistry's reputation and future that needed to be carefully managed. In determining which new synthetic dyes would be harmful as food ingredients, they sought to increase their status as experts as well as effectively legitimizing the use of other synthetic dyes as food additives.

France

Paris was one of the first major cities in continental Europe to set up a municipal laboratory specifically to test for food adulteration. The French laboratory, founded in 1878, primarily served traders, with the regulation of the marketplace as its priority objective.[1] Adulteration concerns in France centered on the falsification of food through, for example, the addition of chemicals, and the

Municipal Chemical Laboratory was set up primarily to detect artificial coloration of wine and other food adulterations.[2]

Stanziani has shown in his extensive research on wine adulteration in France that the use of dyes in wine had been widespread and a normalized practice for centuries. However, the ability of wine producers to make wine using dried imported grapes and the increasing use of chemical products such as synthetic dyes led to a debate in the nineteenth century over what constituted "natural," "agricultural," and "manufactured" food, and how food quality was to be determined.[3] What constituted "natural" and "normal" was a recurrent theme, contested between analysts themselves, as well as producers, retailers, and the public. The new synthetic dyes were never considered natural, unlike many of the other substances previously identified by chemists and the public as adulterants, so examining their reception provides us with a historical insight into how chemical products and scientific innovations are viewed in food production.

As already observed, one of the first chemists to devise tests for detecting aniline dyes in wine was Armand Gautier, who raised particular concern about the use of grenat, a by-product of the manufacture of fuchsine and other aniline dyes used in wine. According to Gautier, grenat had increased in value from being a worthless waste product to being "sold at a remunerative price in consequence of its use for the adulteration of wines." Grenat "consisted of a mixture of mauve aniline, crysotoluidine, fuchsine, and an undetermined body called brown grenat." The reference to grenat's escalating price due to its use in wine suggests that the practice of using the new dyes in wine was widespread by the mid-1870s. As a result, Gautier warned that "aniline should be sought for in all wines found to be adulterated with other substance" and warned chemists to test for arsenic whenever the use of fuchsine was suspected.[4] Gautier suggested that the use of the artificial dyes was successfully passing under the radar of both consumers and public analysts. Like other chemists already discussed, Gautier expressed no concern about the intrinsic safety of the synthetic dyes themselves other than the possibility of arsenic contamination.

In an investigation of food regulations in France between 1851 and 1906, Pierre-Antoine Dessaux demonstrates the complex mediation involved in the understanding and acceptance of chemical additives in food, with the analytical chemists just one group in a "crowded arena" of self-proclaimed experts that included the food producers, retailers, other chemists, and consumers. He concluded that French chemists, while taking an active interest in the use of chemical additives in food, realized that they needed to work in cooperation with the food and drink industry in order to secure any scientific authority.[5]

By the mid-nineteenth century, Paris was one of the most densely populated cities in the world, and public health, *hygiène publique*, including the safety of food and drink, was of growing social and scientific concern.[6] International Congresses of Hygiene and Demography held in Brussels in 1876 and Paris in 1878 led to the establishment of municipal laboratories in the two capital cities and throughout France including Le Havre in 1879, Reims in 1882, Rouen in 1883, Amiens and Saint-Etienne in 1884, and Pau in 1885. The labs were part of a hygiene program whose philosophy, shared by chemists and doctors, was based on the concept that the health of the population could be improved through the use of statistics, safety of food, vaccination, and good housing and education.[7]

The municipal police in France, including the Préfecture de Police in Paris, were responsible for monitoring food and drink. However, laws passed in 1851 provided no standards as to what constituted fraud, resulting in many disputed cases being decided in the courts. Wine, particularly, lay wide open to adulteration.[8] It was not long before French wine producers saw the potential of the brilliant new aniline dyes being sold to textile manufacturers to add color, and profit, to their own products. A series of crop diseases from the mid-nineteenth century had blighted French wine production, encouraging the production of diluted or artificial wines. Meanwhile, concerns began to be raised in both public and scientific arenas about the suitability of the new dyes for consumption. In 1878, following a suggestion by the French organic chemist Jean-Baptiste Dumas, who was a municipal council member in Paris, the Préfecture de Police set up a Municipal Laboratory of Chemistry to test wine for artificial colorings on behalf of buyers and retailers who would pay for the tests. In addition to disclosing the widespread use of artificial coloring in wine production, the laboratory also revealed the extensive use of raisins to make wine and the addition of glucose to mask the dilution of wine with water.

The Municipal Laboratory was well resourced with state-of-the-art facilities and equipment, according to John Muter, a British public analyst who visited in 1885. Muter described a sophisticated setup comprising three laboratories, capable of housing thirty-five workers, plus a private laboratory for the chief analyst with two darkrooms for polarization, a room for gas analysis and vacuum operations, a darkroom for microphotography and spectrum analysis, a room for organic analysis and dialysis, and a distilling room.

While Muter observed that the laboratory was well staffed, he pointed out that the salaries paid to the more senior French municipal chemists compared unfavorably to experienced British analysts, both in the public and private sector. Muter described the salaries of "the chief analyst at £400 per annum, the sub-chief at £300, and 25 assistants whose salaries vary from £220 to £100, together with a staff of '*Inspecteurs experts*', 20 in number, at salaries from £100 to £150." According to Muter "the remuneration of the heads of the departments is low as compared with what chemists of equal standing would require in this country. For example, at Somerset House Dr Bell gets £1,000 and Mr. Bannister £750, per annum; and it is certain that, with us, really competent men could not be got to give up private practice for less than these amounts."[9]

Specialists in wine, or whatever drink or food was under investigation, would assess any sample before it was passed to a chemist to analyze. This suggests that organoleptic skills of taste, smell, and sight remained central to the assessment process even after samples had been passed by the public to police stations and transferred to the Municipal Laboratory.

> The samples are first examined (especially in the case of wine) by experts trained to judge by physical appearances, taste etc and the results having been entered in a special register, they are then divided among the chemists, who are bound to commence the analysis the same day. The chemists are each specialists in some particular article of food or drink, certain men being charged only with the analysis of milk, others of wine, others of fats and oils, and some with microscopic work. This has been found the only way to obtain rapidity of work and control of accuracy. While, however, all are specialists, they yet have sufficient general ability to take part in any branch of work, should a special requirement arise.[10]

Charles Girard, a founder of the Saint-Germain aniline dye manufacturer La Fuchsine, headed the laboratory until 1911, and its chemists worked closely with those employed by the Toxicology Laboratory of Paris. In 1883 Girard's first report as head of the Municipal Laboratory clearly demonstrated the prevalence of adulteration in French food and drink. It met with vitriolic criticism from businessmen, who were concerned about the impact it would have on trade, especially France's valuable wine industry. They argued that criticizing and preventing the use of coal tar dyes was an infringement of liberty, as consumers were entitled to drink aniline-colored water if they wished.[11] Girard defended his laboratory's work in a second report published in 1885, claiming that fighting adulteration with chemistry would benefit French commerce. "Commerce today is quick to change the discoveries of science into instruments of fraud. Falsification, which was formerly based on a few clumsy formulae, has become scientific—and we cannot be successful against it if we don't attack it with weapons equal to its own," he claimed.[12]

Dessaux compares Girard to Harvey Wiley, the head of the US Department of Agriculture's Bureau of Chemistry, describing both as militant campaigners for the consumer, promoting chemistry as a crucial tool to monitor and control the food supply. Certainly both men were keen to build up their own chemical departments and position them at the center of food surveillance, and neither hesitated to publicize what they saw as bad practices in the food industry. Both men also faced determined opposition from the food industry, politicians, and hygienists and found themselves forced to accept compromises.[13]

Businesses, particularly wine producers and traders, argued that the quality of wine could not be determined through science alone, as consumers had different but specific preferences with regard to tastes, smell, and color, while wines varied from region to region, depending on the type of grape grown, and other conditions such as climate, topography, soils, and production methods. Natural variation in food thus made the determination of consistent regulatory standards difficult. Rather than applying prescriptive limits and banning additives or new food practices, continental European countries such as France and Germany increasingly opted for the creation of "norms" for the constitution of certain foodstuffs, standards that often reflected regional differences in production and consumer expectations.[14]

During vigorous legislative debates prior to the passing of food legislation in 1905, Georges Berry, a Paris-based representative of the food and beverage retail industry, warned of the danger of leaving the definition of food items to a committee of bureaucrats and scientists. Such a system, he claimed, would give "the gift of infallibility to chemists whose theories are challenged everyday by their own colleagues" and these new experts "will be given the right to decree formulae with which nature and consumer tastes will have to comply. Truly, it would be rather pleasant, were it not so dangerous for the future, to see nature ruled by infallible chemists."[15] Both industry and politicians also pointed to the inability of chemists to agree in many cases of adulteration. Indeed, as we have seen, consensus within the organic chemical community was far from secure. French chemists such as Girard and Paul Brouardel, a leading forensic chemist and chair of the Comité consultative d'hygiène, argued that chemists should be the final arbiters of food safety. However, other chemists, including Paul Cazeneuve, recognized the lack of consensus within the chemical community. Cazeneuve believed that scientists should not make laws but instead support government and industry to ensure that food and drink were fit for consumption. As a senator from the Rhône, Cazeneuve was a politician as well as a consulting chemist who worked for a variety of businesses, including dye-manufacturing companies. Cazeneuve clearly saw the benefits of close cooperation between chemists and industry.

The growing concern surrounding the use of artificial dyes in wine and conflicting opinion as to their harmful effects led the Paris-based Comité consultative d'hygiène to commission a report from the academic chemist Adolphe Wurtz. Other contributors to the report comprised Henri Fauvel, a chemist at the Municipal Laboratory; the pharmacist Antoine Bussy; Adrien Proust, inspector general of sanitary services and professor of hygiene at the Paris School of Pharmacy; and Georges Bergeron, professor of medicine and forensic physician to the Paris police. The report concluded that most dyes probably were harmless in the small amounts used, but that more experiments were needed to be sure of their safe use in food and drink.[16]

Cazeneuve and his colleague Lépine agreed with the Comité consultative d'hygiène that not enough testing had been performed on the new chemical dyes to properly assess their impact on digestion. As a result of their work, the two chemists recommended that there should be a definitive list of harmful and harmless colors, and that dyes sold for food and drink should be sold under a manufacturer's seal of quality and under official names, and not names made up by retailers. As discussed earlier, most of their experiments showed that the new artificial dyes were not harmful when consumed in small quantities. Food dyes, they argued, should be pure or combined or treated with substances known to be nonpoisonous.[17] As has been shown in previous chapters, much of the concern surrounding aniline dyes lay around their possible contamination with arsenic. In 1885, an ordinance of the police commissioner in Paris had banned the use of fuchsine and other coal tar dyes in food and drink.[18] But French chemists such as Cazeneuve sought a French food law, similar to that enacted by Germany in July 1887, that would clearly separate poisonous colors, containing antimony, arsenic, lead, and mercury, from pure aniline and azo colors. Cazeneuve argued that his and other chemists' experiments showed that many artificial dyes were similar toxicologically to vegetable dyes and other recognized and legitimized food additives such as salt, particularly as the quantity of chemical dyes used was so small. His Lyon colleagues, Lépine and Saturnin Arloing, professor of physiology at Lyon, stated that the use of artificial colors in wine and bouillon was no more harmful than salt or other acceptable additives, while Armand Gautier and Alfred Riche claimed that many artificial dyes were probably no more harmful than natural dyes.[19] As a result of these statements and objections made by the food and drink manufacturers, an 1890 Paris police ordinance included both a list of banned coal tar dyes and a list of coal tar dyes acceptable in food and drink if used in small quantities.[20]

The situation in France highlighted both the uncertainty surrounding what constitutes food improvement and food adulteration, as well as the lack of consensus between chemists at the time. While chemists such as Girard, who were appointed to police the nation's food supply, urged that stricter measures be taken against the new dyes, other chemists, including physiological and organic chemists, believed most of the new dyes were harmless in small amounts, and, in some cases, provided benefits to food production.

In the end, the 1905 French food law was a compromise between all stakeholders, including chemists and the food industry. It focused on the provenance of food, seeking definitions and norms for different products produced from different areas, improving both transparency and regulation of the marketplace, alongside a standardized system of expertise dependent on quantities and measurements defined by chemists. The legislation was the first in a legal process based on the French system of *terroir* which led to the Appelations d'Origine Contrôlées, founded in 1935 by the Ministry of Agriculture as a system of quality control based on place of production.[21] The French system of food regulation and expertise, which was copied by many European countries and was subsequently translated into European Union regulations on food, appreciated the individuality of food produced according to different regional growing conditions, practices, and preferences, and recognized the lack of uniformity in natural food products. While French legislation recognized regional variations in food, the administration and bureaucracy of food monitoring and safety itself became more centralized when in 1906 a ministerial decree created a Service for the Repression of Frauds at the Ministry of Agriculture. This decree brought municipal laboratories throughout France under the control of the Ministry of Agriculture, shifting the balance of bureaucratic control from region to state.[22]

While the food chemist had an important role to play within the new French food regulatory system, food producers maintained the upper hand. Indeed, food regulation in France actually helped to establish and legitimize the presence of the new dyes in food production.

Germany

Another country where differing regional tastes and opposing institutional, scientific, and economic interests converged in the adulteration debate, and in the use of the new chemical dyes in food, was the newly formed nation of Germany. Unsurprisingly in a nation composed of formerly autonomous states with diverse histories and traditions, tastes and food types varied from province to province. New production techniques and novel food ingredients, as well as a changing understanding of physiology and of the importance of different food groups, created confusion and anxiety among consumers and led to calls for more "natural" food products and a desire to stamp out adulteration. However, what counted as illicit versus licit food intervention and ingredients was contested and became the subject of ongoing debate between food producers, retailers, consumers, chemists, and government officials.[23]

In 1871, the year of Germany's unification, the German Reich set up the Imperial Health Office (Kaiserliches Gesundheitsamt) in Berlin. The German historian Vera Hierholzer has termed the Imperial Health Office "a permanent scientific counselling body" and describes the late nineteenth century as a period of an "intertwining" of state and science, with chemists called upon as important witnesses and decision makers in food negotiation.[24] By 1879, a national law on food, which included the banning of toxic dyes, had been introduced in an attempt to establish uniform food control.[25] Based on the British 1875 Sale of Food and Drugs Act, the German legislation aimed to protect the consumer against substitute food ingredients that could reduce the nutritional value of food items and deprive the consumer economically.

Chemists became key components in the state's new food regulatory and monitoring network in Germany, and by 1907 the state's regime included up to an estimated 180 private, state, and university food laboratories.[26] However, there remained considerable regional differences in both implementation and interpretation of the national food laws. In 1879 the Munich-based Institute of Hygiene, set up by Max von Pettenkofer, a former student of Liebig, took eighty thousand food samples; a year earlier, in 1878, a total of only seven samples in Dortmund, fifteen in Köln, and three in Münster had been taken.[27]

Evidence from an investigation conducted by the Imperial Health Office in 1899 to assess the extent of chemical colorings and preservatives used in sausages demonstrates the extensive variation in food practice and monitoring across Germany at the time. Replying to the health office's request for information regarding the use of chemical colorings and preservatives, Berlin's Chemisches Laboratorium Institut für Mikroskopische und Bakterioskopische Untersuchungen noted that most sausages tested contained *Theerfarbstoff* (aniline dyes). Replying to the same survey, however, Offenbach's Chemische Untersuchungsamt reported only one colored sausage in the previous three years, which reportedly came from Frankfurt and was dyed with cochineal, an animal-based dye extracted from beetles. Arthur Forster, a food chemist from the Chemische Untersuchungs-Stelle Plauen, reported that colored sausages had not been encountered in his district since a butcher and his suppliers had been strictly punished for artificially coloring sausages in 1897. However, the Medicinal Bureau in Hamburg provided detailed lists of the different types of dyes used in various sausages, including eosin, rosaline, and other coal tar dyes, while the Hygienisches Institut der Universität Leipzig noted the regular use of azo dyes in sausages. After gathering evidence from public and private laboratories across Germany, the Imperial Health Office confirmed that aniline and azo dyes were regularly used in sausages and many other meat products. The report suggested that the replacement of traditionally used dyes such as cochineal with the new chemical dyes had become normalized with the "opportunity provided and incentive created to find alternative colours" as a result of the German food law of July 5, 1887. This law had banned several, mainly metallic, dyes known to be harmful, thus driving producers to search for replacement dyes and effectively legitimizing any dye not containing substances, such as arsenic, which were banned by the act.[28]

As a newly unified country, Germany faced particular problems in establishing "norms" in food production, making the issue of deception and adulteration problematic. Food producers and retailers and consumers in different states had differing opinions as to what constituted legitimate or dishonest ingredients, while chemists, nutritionists, regulators, and the food industry could not agree on allowable limits of certain ingredients, particularly chemical dyes and preservatives. While the growing body of chemists, nutritionists, and physiologists in Germany argued over how to assess the dyes scientifically, consumers from Berlin to Bavaria and Baden varied in their assessment of taste, color, smell, and texture of food products. Assessing to what extent chemical additives should be used in food production depended on consumers' sensory perceptions of what standard colors and flavors should be as much as the commercial constraints of profit-making food producers and retailers, the opinions of chemists or the political constraints imposed by the need to feed a growing urban population economically and safely.[29]

Germany's food legislation had left open the definition of what constituted adulteration, creating a broad legislative and regulatory framework into which chemists, keen to professionalize themselves and raise their political status, could expand. Germany was a leading educator of chemists, and the numbers of chemists employed in Germany's workforce expanded dramatically in the last quarter of the nineteenth century. The expanding numbers of chemists involved with food chemistry, technology, and nutritional science began to form and publish their own opinions and guidelines to monitor and regulate food.[30] During the 1890s, the Free Association of German Food Chemists (Freie Vereinigung Deutscher Nahrungsmittelchemiker), Germany's equivalent to Britain's public analysts and France's municipal chemists, began to compile a series of uniform norms and guidelines for food content and preparation. These standards were adopted by government and published by the Imperial Health Office.[31] However, as food became more industrialized, commercialized, and subject to political control, food producers and retailers as well as consumers also set up their own lobbying groups. In 1901 the Bund Deutscher Nahrungsmittel-Fabrikanten und Händler (Union of German Food Producers and Retailers) was formed to represent food producers and retailers in a food market that they felt was increasingly being regulated by the chemists and politicians with insufficient consideration of the food industry itself.[32] In 1905 the union published its own book on food norms and guidelines, the *Deutsches Nahrungsmittelbuch.*

Disputes between the food industry and state and privately funded independent chemists were frequent and vociferous. From 1883, after the food industry complained that chemical analysts were not being sufficiently pragmatic in their approach to food adulteration, the Reichs-Justiz Minister ruled that the country's courts had to consult experts from the food industry as well as independent and state food chemists before determining whether foodstuffs had been adulterated. The ability of the food producers to undermine the expertise of the food chemists in court, and to argue that food was not adulterated if it was safe, economic, and "normal," angered food chemists, who believed that food norms and ingredients needed tighter regulation, particularly with regard to chemical additives.[33] Professor Carl Neufeld, head of the Munich Research Laboratory, objected to what he viewed as a sidelining of the expertise of food analysts, arguing that the food chemist was not a mere analyzer but a key contributor to any judgment.[34] At the XIVth International Congress for Demography and Hygiene in Berlin in 1907, Josef König, head of the Münster Agricultural Research Station, declared that if the habits of industry were treated with impunity, then no grounds could ever be invoked for any claim of falsification or adulteration.[35] Neufeld and König recognized that chemical colorings were used widely in food production and that most were safe in small quantities. However, both argued that such additives should not be used to deceive consumers and should not be consumed in large quantities.[36] Neufeld, König, and other food chemists continued to press their case for more independent analysis and to reproach the food industry for using professional advisers and skillful advocacy to flout regulations. In 1910, the German government passed a new law that emphasized the importance of the independent food chemists' expertise, while allowing commercial expert witnesses representing the food industry to appear only in doubtful cases and where they could prove their professionalism and impartiality.[37] Consulting chemists were not strangers to courtrooms in the nineteenth century, and were often employed as expert witnesses. However, conflicting evidence presented by chemists and other so-called expert witnesses in court often became the subject of much controversy, as already explored in the case of British analysts.

Both the independent food chemists and the German food industry had set up their own versions of food norms, and often disagreed as to what constituted a norm, particularly when it came to the use of chemical preservatives and colorings. Food companies claimed that the new chemical preservatives were often preferable, as they had less impact on flavor and were no more harmful than existing and long-standing methods of preserving such as salting and smoking. They argued that many new food products, such as margarine, had no longstanding norms and claimed to use artificial coloring to meet consumer preferences, which often differed from region to region. Food producers also pointed out that even household cookbooks recommended the use of chemical preservatives such as salicylic acid and artificial colorings.[38] While public analysts sought to ban the use of chemical colorings, or at least limit their use, the *Deutsche Nahrungsmittelbuch*, the food industry's handbook, declared that "the coloration of a foodstuff for nutritional physiological reasons has to be seen as a material improvement, and, purely from a physiological perspective, to be treated with equal importance to seasoning." Pavlov's experiments had shown that food color was part of the psychology of food, the food industry claimed, arguing that, as the color of food encouraged people to eat it, the dyeing of food to improve its attractiveness should be considered an improvement to food.[39]

Like the organic chemists and public analysts, the German physiological and hygiene chemists also had a stake in safeguarding the reputation of chemistry and its products. Theodor Weyl and his fellow German chemists understood the importance of the growing chemical industry to the German economy. They recognized the advantage of extending the range of chemical products, both for use in science and society, while acknowledging the pervasive manner in which the new chemical substances were infiltrating everyday life and the need to reassure the public. Hygienic and physiological chemists identified toxicity testing of the new dyes as a way of expanding their repertoire and status, in a similar manner to the way in which public analysts had identified food adulteration as a means to increase their position and professional work earlier in the century.

In the preface to Weyl's publication on coal tar dyes, his colleague, Eugen Sell, observed that the coal tar color industry of Germany had "conquered the world" and that as a result of the "steady work" of the German chemical industry, more and more products were entering the market. Sell underlined the important role of hygienic chemists in the project, suggesting that it was only through their expertise that the safety of chemical dyes as food colorings could be secured.[40]

Eosin—a German case study

The material, practical, and social complexities facing German chemists in determining the appropriateness of using the new dyes in food is illustrated in a dispute over the use in animal fodder of eosin, a red dye created by the action of bromine on fluorescein.[41] The controversy demonstrates how politics and commerce played as central a role as science in debates surrounding the appropriate use of dyes in food. The case also shows that consensus still had not been reached on toxicity testing of chemical dyes on animals by the second decade of the twentieth century.

The discovery of eosin in the early 1870s was a good example of successful collaboration between academic and industrial chemists in Germany. Eosin was one of the first phthalein dyes, made by condensation of phthalic acid, extracted from the coal tar hydrocarbon naphthalene. The discovery of eosin Y, a disodium salt of tetrabromofluorescein, and eosin B, a sodium or ammonium salt of the dibromo derivative of dinitrofluorescein, was the result of collaboration between BASF's leading chemist Heinrich Caro and Adolf von Baeyer, then professor at the University of Strasbourg.[42] Eosin dyes became more extensively available after AGFA's Carl Martius purchased a sample of BASF's product and gave it to his consultant Hofmann at the University of Berlin, who published the formula and details of the reactions, revealing what had previously been an industrial secret.[43] One wonders whether eosin became the target of public concern precisely because, unlike many of the hundreds of unknown and undetectable dyes being used at the time in food, the secrets of its production and constitution were more widely visible in the public domain.

In 1909, several newspapers including the *Berliner Local-Anzeiger Morgenblatt* raised concerns about cheap animal feed, much of it imported, being colored with eosin.[44] As illustrated in other chapters, the linking of foreign food to adulteration and toxicity is a recurring refrain. The article alluded to reports in the "corridors of the Reichstag" that eosin had been tested on animals and found to have disturbed stomachs and intestines. The widespread reporting of these concerns led to a series of written exchanges and reports between various government departments and scientific committees, including the Interior Ministry (Staatsministerium des Innern), the Imperial Health Office (Kaiserliches Gesundheitsamt), the Royal Prussian Medical Committee of Experts (Königliche Preußische Wissenschaftliche Deputation für das Medizinalwesen), the State Crop Agency, and even the Treasury, which was concerned about the possible economic impact of the destruction of vast quantities of animal fodder.

On May 14, 1910, Franz Bumm, the president of the Imperial Health Office, sent a letter to the interior minister, rebutting the newspaper reports and enclosing several reports of experiments that demonstrated that eosin was not harmful as a food dye. These included a June 1904 report of experiments commissioned by the Imperial Medical Committee of Experts, which showed that feeding rabbits with eosin-colored barley in restricted amounts was not harmful, and the results of a series of experiments conducted by Imperial Health Office chemists. The office's president also noted that the French Conseil Supérieur d'Hygiène Publique and the Académie de Médicine in Paris had agreed to legislation in August 1908 that permitted the use of eosin in France as a color in liquors and syrups.[45]

An enclosed publication by the Imperial Health Office member and state chemist E. Rost outlined the experiments undertaken by the office. These included a series of experiments on rabbits, dogs, cats, fish, and even humans. As a result of the experiments, Rost concluded that eosin was generally not absorbed into the body and only a small part remained in the organism. Rost found that eosin produced no physiological disturbances in most of the animals tested, that it was not possible for a dog to die through eosin in the stomach, and that the dye produced no disturbance of the nervous system. In a characteristic reference to the dangers of foreign produce, Rost claimed that eosin could only be harmful in very large doses, as might occur in foreign fodder.[46]

However, the controversy continued after the outbreak of war in Europe. On October 9, 1915, Fritz Schanz, a Dresden eye specialist, wrote to the Imperial Health Office claiming that government regulation allowing the use of eosin was dangerous for animals and people, as eosin could be fatal if ingested by animals exposed to sunlight. Schanz claimed that his work had shown that intense light changed the proteins in blood and that this change was heightened when certain materials, known as photocatalysts, were also present in the blood. Schanz claimed that eosin was such a material. Enclosing copies of his treatises, *Sonnenstich-Hitzschlag* (Sunstroke) and *Die Wirkungen des Lichtes auf die lebenden Organismen* (The effect of light on living organisms), Schanz warned that even in small doses, eosin could harm bodies exposed to sunlight and also the humans who consumed them. Moreover, he claimed that the experiments conducted by the Imperial Health Office were undertaken on animals confined to stalls and therefore excluded the effect of light.[47]

Newspapers picked up on the accusations, publishing articles about the impact of sunlight on eosin's toxicity, and questioning why, if lemonade and fruit drinks carried labels claiming that they did not contain eosin, the chemical was still fed to animals. Rost, outraged at the continued criticism from Schanz, particularly his accusations that the Imperial Health Office experiments were flawed in their premise, published his own articles in the trade and professional press, pointing out that no problems had been found despite extensive testing. He also claimed that in his opinion, and that of all literature other than Schanz's work, eosin was harmless and added that there were more significant things to worry about in a time of war, such as incidences of pellagra among German troops.[48]

On October 11, 1915, the German State Crop Agency (Reichsgetreidestelle Geschäftsableitung Gesellschaft) wrote to the Interior Ministry urging it to resolve the matter, fearing that Schanz's accusation might result in the wholesale destruction of fodder containing even small amounts of eosin. In a reply dated November 9, 1915, to the director of the State Crop Agency, the Imperial Health Office president described Schanz as well regarded by the public in Dresden and as one of the best eye doctors, with a private practice and a good reputation as an eye surgeon and a member of the Saxony Health Board. Schanz had developed glasses for reducing the effects of ultraviolet rays on eyes and had performed physiological tests on the effects of sunlight on the protein of the eyes with fluorescein substances. The president observed that Schanz was largely self-taught and not a trained chemist, as he often admitted himself. Although he censured Schanz for criticizing the Imperial Health Office, he conceded that Schanz's complaints came from a concern over people's welfare. The president noted that he had reluctantly submitted a resolution to the Interior Ministry, a step he had never taken before, and asked whether, should the resolution go back to the state health department for peer review, his exchange about Schanz could be kept confidential.[49]

A special supplement in one of Germany's leading weekly medical publications was devoted to the squabble between Schanz and Rost, with each scientist given space to put forward his view. Rost claimed that Schanz's "groundless claims were unsettling the civilian population in this serious time" and that Schanz should "desist from his speculations in the interest of the Fatherland." Schanz, meanwhile, claimed that since 1909 animals throughout Germany, including pets, had shown symptoms of eosin poisoning. While the storm of protests that had resulted had prompted the government to commission more tests, this research had begun, Schanz claimed, on 3 January 1910, when light levels were low, invalidating the experiments. Dr Bumm, the Imperial Health Office President, wrote to the State Crop Agency in March 1916 with reference to both articles, explaining that he had asked Schanz to provide evidence for his claims. The dispute continued after the war ended, at which point Schanz, to no avail, sought government funds to prove his claims were justified.[50]

This episode encapsulates many of the issues involved in reaching consensus. The two scientists attempted to prove their separate hypotheses using experimental evidence in a field where experimental and theoretical consensus had not yet been secured. Indeed, as already described, experimental consensus surrounding the toxicology of the coal tar dyes remained, and still remains, elusive. The two scientists were operating within different disciplinary frameworks and within separate social and professional settings. When no agreement could be reached

on establishing the facts of the case, including the experimental parameters and conditions necessary to obtain them, the scientists fell back
on issues of personal credibility, each trying to secure support for his
hypotheses through the media and by lobbying peers, the public, and
politicians. Social, economic, and political aspects played as large a part
as the science in a situation that was never securely resolved.

8

The US government acts against chemical dyes in food

While the US lagged behind Europe in the introduction of
food legislation, many twentieth-century historians have
portrayed the US 1906 Pure Food Act as the blueprint for
modern food regulation, and its architect, state chemist
Harvey W. Wiley, as a consumer champion who stood up
to big business in the pursuit of natural food.[1] Wiley, like
chemists in other countries, identified food safety as an
area where analytical chemists could exert their authority
and improve the public food supply. But, in common with
his peers in Europe, Wiley soon realized that he needed
to work with, rather than against, food and chemical
manufacturers.

By the end of the nineteenth century, concern over
both adulteration and industrialization of food was rife
across America, with demand for reform coming from
politicians, food chemists, large food manufacturers, representatives from rural and urban communities, and the
middle classes, particularly women. The resulting legislation was, therefore, an attempt to marry a wide variety of
concerns, interests, and viewpoints. While Wiley was concerned about the health of the nation, he also understood
the concerns of industry and the need to build a coalition
between businesses, consumers, and politicians. Historians have demonstrated how the larger US food producers
successfully used regulation to their commercial advan-

tage and how Wiley himself garnered big business support as well as backing from consumer campaign groups (especially women's groups), state regulators, and physicians to help get his measures passed, expand government control, and increase the status of analytical chemists and the US Bureau of Chemistry.[2] Among Wiley's principal tactics was to try to convince food suppliers that the American public did not want food preserved and colored with manmade chemicals and that banning such "adulteration" would benefit honest food manufacturers at the expense of those prepared to deceive and defraud consumers.

> I told the canners exactly the truth. . . . that they, instead of helping their business by claiming right to put a coloring matter in their goods, were driving hundreds of thousands of American citizens away from the stuff.[3]

US businesses, however, like those in France and other countries, complained that there was no consensus among chemists about whether, and which, chemicals were safe to add to food and argued that food producers should not be under the control of a government chemist. Ketchup manufacturer Walter H. Williams, president of the Walter H. Williams Company of Detroit, argued in congressional hearings:

> I don't think this committee ought to recommend any legislation which will give one man the absolute power to say what the manufacturers of this country shall do and what they shall not do. There is a difference of opinion as to what is injurious and what is not injurious. We can show that the very best scientific thought in this country will differ from the present Bureau of Chemistry. Now, gentlemen, do not understand for a moment that I am attacking Dr Wiley or the Bureau of Chemistry or the Department of Agriculture. I am simply pointing out, or trying to point out, the principle of this bill. The principle is wrong.[4]

Wiley for his part claimed that he did not oppose the use of chemical colorings and preservatives in food entirely, pointing to some circumstances where chemical additives were needed to lengthen shelf life in inaccessible places like at sea and in countries unable to produce food for their entire population "such as England." He also believed that if the consumer wished to buy food colored or preserved with chemicals they should not be denied that preference. His arguments against the increasing use of chemicals in food were, he claimed, applied "to the usual conditions which obtain in this country and especially to the apparent fact that the great majority of our people seem to prefer their food untreated with noncondimental preservatives."[5]

Wiley's statement recognized the distinction made by consumers between the new industrially prepared chemicals being increasingly used in food and traditional preservatives such as salt, sugar, and smoke curing, and colorings extracted from fruit, spice, and vegetables that had been used in the preparation of food for centuries. As in other countries, the introduction of chemical additives as colorings or preservatives proved to be a particularly controversial area where the views of scientists, the food industry, and consumers conflicted both with one another and with those of other interested parties. These new "scientific" substances provided benefits for everyone, extending shelf life and protecting against germ contamination while decreasing the costs and increasing the availability of food. However, concerns about the possible, but unknown, physical risks of consuming increasing amounts of synthetically produced chemicals and the ability of food producers to use the chemicals to deceive the consumer began to increase. As in Britain, one food product that proved to be especially controversial in the use of artificial colors was oleomargarine.

Many countries legislated against this novel food product, but some of the most concerted and powerful dairy lobbying occurred in the US, particularly dairy farming states. A vivid insight into the fears of the dairy industry at the time can be seen in an 1886 document published to persuade legislators to act against oleomargarine. The publication complained about the power, influence, and financial might of the oleomargarine producers in Europe and their effect on the American dairy industry and exports. The following quote gives a flavor of the antiforeign arguments adopted by the US dairy industry in its crusade against margarine. It also alludes to conflicting evidence and theories put forward by scientists, a complaint regularly used by opposing factions as a way of discrediting science:

> OLEOMARGARINE, the basis of all the frauds in butter, is the outcome of an ingenious Frenchman's notion that the butter diffused through the milk of the cow is due to the absorption of the animal's fat. Taking some minced beef suet, a few fresh sheeps' stomachs cut into small pieces, a little carbonate of potash and some water, this Frenchman, Hippolyte Mege by name, subjected the mixture to a heat of 113 degrees Fahrenheit; and so, by the action of pepsin in the sheeps' stomachs, separated

the fat from the other tissues. By hydraulic pressure this fat was again separated into stearine and margarine; and putting ten pounds of the latter into a churn with four pints of milk, three pints of water, a little annatto, Mege succeeded in turning out a compound sufficiently like butter to pass for that article, its only lack being the golden yellow color that characterizes all good butter. Whether he had produced a deleterious stuff containing the germs of disease and of all manner of loathsome parasites, as one set of scientific experts pronounced, or something far more wholesome than half the butter in the market, as another set emphatically declared, was of little moment to the discoverer, so long as the thing was likely to prove profitable. He patented his process, and found no difficulty in selling rights to handle it in France, England, Holland, Germany, and the United States.[6]

The appeals for a return to natural food as a way to counter the degeneration of society and the individual was invoked by participants in the public debate over margarine.[7] Dairy farmers appealed to social hygiene issues, which were widespread among middle-class politicians at the time, to enlist political support to restrict the import and sale of margarine by emphasizing its social menace. The dairy industry claimed that margarine caused dyspepsia and other illnesses, and contained putrid beef, dead horses, mad dogs, drowned sheep, and workmen's toenails, and dairy lobbyists pressed politicians to legislate against the new impostor.[8] Instead of being viewed as an additional and innovative food product offering an economical choice to the consumer, margarine was here portrayed as a case of fraud, an example of industrialized and unnatural food that became a scapegoat for the ills of the dairy industry and for society as a whole. Mechanized margarine factories were represented as a threat to the pastoral idyll of the dairy farmer.[9]

Ironically, at the same time that margarine was portrayed as the industrialized impostor, the manufacture of butter itself was becoming increasingly industrialized. Butter manufacturers were adding water or vegetable fat to their own product in order to boost their profits. Retailers further exacerbated the deception by disguising inferior products, from both margarine and dairy producers, as premium butter. However, it was margarine that became the adulteration scapegoat, as the butter industry, with its preexisting power base in several influential US states, proved to be one of the earliest and most effective special interest lobbying groups in US political history.[10] The anthropologist Mary Douglas argued that when "ritually significant foods take on moral sanctity, their rivals come to be defined as stigmatized impostors carrying moral dangers."[11] However, "ritually significant foods" themselves are constructs of society, and in this situation the "tradition" of butter became invented for commercial purposes by the dairy industry.[12]

As a result of the dairy industry's successful lobbying, the state of Missouri prohibited the making and selling of margarine in 1881. Other dairy-producing states, including Maine, Michigan, Minnesota, Pennsylvania, and Wisconsin, followed suit in 1885. Penalties included fines of up to $1,000 and a prison term of up to a year. Federal law restricting the manufacture, import, or retail of margarine was passed in 1886. The 1886 Oleomargarine Act imposed a tax of two cents a pound on domestic margarine and fifteen cents a pound on imported margarine. In addition, margarine makers were required to pay a license fee of $600 per annum, with annual license fees for wholesalers of $480 and for retailers of $48. Christopher Burns claims that oleomargarine was regarded as taxable because of its low cultural status, in common with the other taxed commodities of alcohol and tobacco. He argues that such taxes were "moral taxes designed to regulate what people consume" and that the 1886 act was effectively a "stigmatization" of oleomargarine by the US government.[13]

In the early days of margarine, the toxicity of artificial dyes was not a great worry. The major concern was fraudulent attempts to disguise margarine as butter by coloring it yellow. It passed unnoticed that much butter was also colored. As a result, campaigners sought to regulate the coloring of margarine even in states where the sale and manufacture of the new product were not banned. In 1884, New York forbade the coloring of margarine to make it look like butter, and the New York State dairy commissioner requested that quicker and easier methods be found for the detection of artificial coloring substances in the product. American chemists such as Edward Martin and Elwyn Waller of the School of Mines, Columbia College, in New York, and Albert R. Leeds, professor of chemistry at Sevens Institute, Hoboken, New Jersey, obliged. A table produced by Leeds showed how butter and imitation butter colored with different dyes including annatto, turmeric, saffron, carrot, marigold, aniline yellow, Martius yellow, and Victoria yellow produced different colors when treated with various acids including sulphuric acid, nitric acid, and hydrochloric acid.[14]

By 1900, thirty four states had banned yellow margarine, with five requiring all margarine to be colored pink so that consumers could tell the product was margarine and not butter. Similar restrictions on margarine continued to be imposed in Canada and Australia until well into the twentieth century.[15] Clearly governments in this situation were not concerned about the harmfulness of using artificial coloring, for they were requiring that margarine be deliberately and artificially colored differently from butter in order to protect the consumer from deception. In 1898, in *Collins v. New Hampshire*, the Supreme Court rejected the pink margarine legislation enacted by New Hampshire, ruling that it "necessitates and provides for adulteration" and was prejudicial against margarine.[16] These examples demonstrate the tension between concerns about artificial coloring as a health risk and its use as an instrument for other purposes.

Other countries introduced legislation to differing degrees and with various motives. In the Netherlands, for example, laws to ensure the consistent quality of both butter and margarine were designed principally to restore the nation's reputation as a reliable exporter of both products. This arose as competition increased from other countries such as Australia and New Zealand, which employed new and scientific methods in their butter-making as compared with the Netherlands' traditional and empirical processes. As a leading margarine exporter, the Netherlands was also keen to overturn margarine's international reputation as an "adulterated" butter by policing quality.[17] In most countries, particularly the US, margarine remained subject to high taxes and continuing restrictions and prohibitions until the middle of the twentieth century.

The margarine case shows how the main anxiety surrounding the use of artificial colorings originally centered on their potential to deceive, and thereby cheat, the consumer. This ability to defraud the consumer was of major concern to producers and retailers who feared competition. As a result, much of the driving force for food regulation came from the food industry itself, particularly the large producers and retailers who would benefit from a more regulated marketplace, or, in the case of the dairy farmers, suppliers whose longstanding products became threatened by the emergence of new food products or ingredients.

As new industrial food practices and chemical ingredients continued to inflame the growing and diverse "pure food" lobby, Wiley argued that chemistry itself was the key to policing food and enforcing food standards. In the US, as in Europe, chemistry had become increasingly professionalized over the nineteenth century, with different chemists specializing in different areas. American chemical engineers working alongside engineers in production and manufacturing were increasingly better paid than analytical chemists.[18] In 1884, the Association of Official Agricultural Chemists, of which Wiley became president in 1886, was founded to advance the professional status of analytical chemists in agriculture, to build closer ties to government in order to expand their standing and role in society and reduce adulteration of fertilizers.[19] Wiley himself had studied chemistry after taking a medical degree in Indiana. He then traveled to Germany to learn more chemistry like so many of his European and American peers in the profession. During his time in Germany, he attended Hofmann's lectures and worked with Virchow in Berlin and then studied the chemistry of foods with Eugen Sell, then the chemist of the German Imperial Health Office.[20]

As the head of the Department of Agriculture's Bureau of Chemistry in 1883–1903, Wiley was in an appropriate position to help raise the status of analytical chemists, and his crusade to improve food analysis and protection through the pure food movement was an excellent opportunity to achieve this. In 1903, the Bureau of Chemistry started publishing purity standards for food products, put together by the National Committee of Food Standards, a group composed of representatives from the Department of Agriculture, the Association of Official Agricultural Chemists, and the Association of State Dairy and Food Departments as well as other experts.[21] The Bureau of Chemistry then began work on regulating food additives, particularly preservatives and dyestuffs. As is clear from the situation with margarine, states across the US had different and often conflicting regulations regarding the use of coal tar dyes and many food producers and retailers wanted a uniform approach.

Regulating the use of chemical dyes in food in the US

However, specific regulations about dyes were not included in the original 1906 Pure Food Act. Wiley realized that it would take time to determine which of the many hundreds of dyes being distributed in the market were safe to use. To oversee the analysis, he appointed Bernhard Hesse, an American chemist who had worked for BASF for several years before becoming an independent consultant in 1905. Born in Michigan in 1869, Hesse was the third person awarded a Ph.D. in chemistry from the University of Chicago, in 1896. He had been employed in BASF's German headquarters before returning to New York as the company's technical expert in patent litigation; he reputedly "had been privy not only to many of Badische's trade secrets, but also to matters of company and cartel policy."[22]

By September 1906, Hesse had collected about ninety samples of coal tar dyes recommended by dye distributors for use in food, several of which he believed were harmful. However, by February 1907 Hesse had only managed to test the red samples, warning Wiley that testing might have to continue after regulations had been developed, because each test was so time-consuming. As a compromise, due to time pressures, Hesse suggested producing a short list of dyes that could be "considered harmless and permitted for use." He observed that lawmakers in some other countries had banned specific colors proven to be harmful but pointed out that such measures had not been particularly effective, as they effectively allowed the unrestricted use of all colors not banned, including all coal tar dyes not already examined and newly created colors.[23]

The legislation adopted in Europe, which effectively permitted the use of any dye not specifically banned by government, contrasted both with the approach favored by the US and the laissez-faire attitude taken in Britain, demonstrating how health policies and priorities differed significantly by region and culture. However, the more precautionary approach of the US was almost impossible to implement, for by the turn of the century, there were already too many dyes on the market to police effectively. Hesse estimated that there were 695 different colors manufactured by some thirty-seven separate companies on the world market, making it

> physically impossible to go to every user of coal-tar colors in food products in the United States and obtain specimens of the coal-tar colors so employed: this would be impracticable not only because of the large number of such users, and their wide geographical distribution, but also because they often do not know what they are using, and further because of a reluctance, undoubtedly to be encountered among many, to disclose the nature of the products employed. This is rendered more than likely by the attitude of some of the makers of coal-tar colours.[24]

Despite a huge amount of effort and resources being expended, it was impossible to test every dye to determine its safety, or even to obtain a complete set of samples of test material. Under these circumstances, the US government decided to ban all dyes, whether toxic or not, apart from a few that could be certified as being safe to use. To determine which dyes should be permitted, Hesse consulted various toxicological studies, including those undertaken by the National Confectioners Association and by German chemists Theodor Weyl and Sigmund Fraenkel. Hesse also visited US-based dye manufacturers, including Hartford-based Schoelkopf, Hanna of Buffalo, Newark-based Heller, Merz and Co. of Hartford, and the US subsidiaries of BASF and Hoechst, as well as the specialist dye importers and suppliers to the food industry H. Kohnstamm and Co. and H. Lieber and Co.[25] Hesse surveyed thirteen manufacturers worldwide and four importers, noting that five of them "or 29 per cent have not found it in their interest to contribute either specimens or information." Meanwhile, among the samples provided, there was "very little unanimity among the different concerns furnishing coal-tar colors for use in food products as to which of their products are desirable, necessary, or suitable for such use."[26]

From more than two hundred samples supplied, Hesse whittled down seven colors he determined were suitable for use in food, if guaranteed to be manufactured in a pure form, based on the toxicological studies conducted by Weyl and others. The seven dyes comprised two yellows (napthol yellow S and tartrazine); four reds (carmoisine, rhodamine B, amaranth, and erythrosine); and light green SF bluish. However, less than a week later, Hesse had changed his mind, and, after another process of selection and elimination, chose a different list of seven permitted dyes, only three of which overlapped with the previous list. The final list, eventually approved by Washington and included in Food Inspection Decision 76, "Dyes, Chemicals and Preservatives," issued on June 18, 1907, consisted of red (amaranth); scarlet (ponceau 3R); bluish red (erythrosine); orange (orange 1); yellow (napthol yellow S); green (light green SF yellowish); and blue (indigo disulphonic acid).[27] FID 76 stated that:

> The use of any dye, harmless or otherwise, to color or stain a food in a manner whereby damage or inferiority is concealed is specifically prohibited by law. The use in food for any purpose of any mineral dye or any coal-tar dye, except those coal-tar dyes hereinafter listed, will be grounds for prosecution. Pending further investigations now underway and the announcement thereof, the coal-tar dyes hereinafter named, made specifically for use in foods, and which bear a guarantee from the manufacturer that they are free from subsidiary products and represent the actual substance the name of which they bear, may be used in foods. In every case a certificate that the dye in question has

been tested by competent experts and found to be free from harmful constituents must be filed with the Secretary of Agriculture and approved by him.[28]

Many dye manufacturers, particularly the large European manufacturers, claimed that it was impossible to produce dyes to the standards of quality and purity expected by the US government and opted no longer to market food dyes. For their part, food manufacturers complained that the choice of just seven dyes was too limiting and that none of the selected dyes were suitable for oils or fats. The food industry argued that it was up to the government's chemists to prove the dyes were harmful. As a result of lobbying by both the food industry and some dye manufacturers, the Board of Food and Drug Inspection ruled that the Bureau of Chemistry had to prove a dye was harmful before being able to ban its use. This move angered both the Bureau of Chemistry and the US food dye manufacturers, who had worked for several years to improve the quality and safety of food dyes. A compromise was reached with the introduction of Food Inspection Decision 117 in May 1910, which *recommended*, rather than mandated, the use of certified colors. However, the decision was worded in such a way that many assumed the use of uncertified dyes was banned. US suppliers who specialized in food dyes deliberately encouraged this misconception by intensive advertising and promotion. Meanwhile, the states of Idaho, Illinois, Iowa, and Nevada went further and mandated that only government-certified dyes could be used as food colorings.[29]

By creating the concept of "permitted" chemical dyes, the US legislation effectively opened up a new legitimized and specialized market for food dyes, one eagerly embraced by the US chemical companies as an opportunity to differentiate themselves from their bigger European dye manufacturing competitors. In retrospect, however, commentators have argued that the seven colors recommended as permitted colors in 1907 "later turned out to be comparatively poor choices." Indeed, Melanie Miller claimed that by 1987 three of the seven permitted dyes were considered to be "possibly harmful" and one to be "harmful."[30]

Despite being regarded as the architect of a major piece of consumer legislation, Wiley always believed that his attempts to purify American food from unnecessary chemical additives had been hindered by vested political and business interests.[31] Wiley fervently believed that nature provided the best source of food for humans and that synthesized chemicals should be avoided as much as possible. He argued that while chemical additives may be included in food in small doses, a lifetime of eating food with such additives put excessive strain on internal organs such as the liver and kidneys.

> I believe this with all my heart, that when man eats a normal food normally the length of human life will be greatly extended. This is what I believe. But if we consume abnormal food abnormally we shall lessen the length of human life.[32]

Aside from health considerations, Wiley considered that chemical dyes in food, particularly, were used primarily to deceive the public and could "not see any reason for legalizing their use."[33]

On March 15, 1912, Wiley resigned as chief government chemist, believing that his department's efforts to stamp our food adulteration were being stifled and his own authority undermined by politicians and vested interests.[34] Within two weeks he became director of food and sanitation for the Good Housekeeping Research Institute, where he remained, testing consumer products and campaigning for pure food, for eighteen years. "Coming out of a field in which all the foremen had been my enemies I entered a garden of activity in which all the head men were my friends. In this favourable environment I have had unrestricted opportunity to carry on my battle for pure food, finding no enemy to stab me in the back."

In a self-penned history of his career written seventeen years after his resignation, Wiley argued that among the regulations he had passed that should be repealed were the three that introduced the concept of permitted food colors.

Conclusion

Chemical dyes synthesized from coal tar were among the first totally man-made, industrially produced chemical substances to be added to our food. Today, thousands of different types of man-made chemicals, hormones, and enzymes are added to our food to ensure longer shelf life, better flavor and appearance, and affordability. Colorings, flavorings, thickeners, antioxidants, sweeteners, and solvents are among a vast range of synthetic additives legally described as food ingredients.[1] Finding out why, and how, the first group of completely new chemicals became accepted and legitimized as food additives helps us to understand the industrial transformation of food and our relationship with it.

These days synthetic chemicals, formerly the cornerstone of nineteenth- and twentieth-century industrial progress, are fast becoming science's bogeyman. These man-made substances are now being linked to chronic diseases and illnesses from asthma to cancer. The huge and rapid proliferation of man-made chemicals, whether in food, clothes, or pesticides, is accused of disrupting ecosystems and DNA, causing global pollution and even contributing to climate change.[2]

The first class of mass-produced synthetic chemicals was able to enter daily life in unpredicted ways, revealing the contested position of scientists in their efforts to manage their use once they entered the marketplace and the home. Nineteenth-century chemists had to contend with other sources of authority, and this problematized their attempts to present themselves as experts in public. Exploring the rise of scientific experts, public health legislation, and an early example of consumer risk management through the lens of these man-made food additives highlights the difficulties of managing new science and technology as it is applied in everyday life.

Despite being brilliantly colored and increasingly ubiquitous, these new substances remained almost impossible to detect for decades. Although created by chemists, the detection and evaluation of aniline and azo dyes in food represented a failure in analytical chemistry. New strategies had to be devised, with chemists from a variety of backgrounds and disciplines, and with different aims and objectives, drawing on a diverse range of repertoires of testing. Understanding these new substances involved a complex set of interacting networks of people and practices, operating both locally and internationally. For many analysts, the enigmatic coal tar dyes in food were substances that proved too problematic to hang their expertise and credibility on because to attempt to determine safe and acceptable thresholds of consumption for the multitude of different dyes was beyond their technical capability.

Examining the work of analytical chemists both inside and outside their laboratories has demonstrated the complexities of achieving consensus in the creation of universally agreed-upon tests and the social negotiation involved in determining which experiments and "facts" are applicable, confirming earlier work by scholars in the history and sociology of science.[3] Chemists employed theory and practice from across a wide range of chemical traditions and developments, combining both synthesis and analysis, in their attempts to detect the use of coal tar dyes in food and to understand and assess the impact they had on those consuming them.[4]

Historians of science have claimed for several decades that science is socially and economically driven and that science in turn affects the social domain. Looking at the synthetic dyes as an exercise in how scientists understand and monitor an industrially produced scientific product once it becomes widely consumed affirms these claims. It also shows how important it is for historians to break down the distinctions often made within the history of chemistry between analytical and synthetic chemistry, pure and applied chemistry, and academic, industrial, and consulting chemists.

Additives or adulterants

By examining a specific synthesized substance, and the changes in its application and how it is perceived, we can gain important insights into those manipulating it. Exploring the passage of new chemical products created in the laboratory, manufactured in a factory, and incorporated into food reveals the mediation necessary to accommodate, relate to, and understand new substances. This is an investigation of undetectable substances that inadvertently became hidden, yet intrinsic, parts of everyday life.

How chemical products and innovations were viewed in food production by the press, public analysts, hygienists, food producers, the state, and the consuming public provides an intriguing insight into the legitimation of science and, more specifically, the legitimation of manmade chemicals as food ingredients.

The presence of synthetic dyes in food was a contested issue from the outset among the chemical analysts involved in food monitoring and food production as well as researchers involved in physiology, histology, and organic chemistry. Many food chemists believed that the new dyes were preferable to the existing mineral and vegetable dyes, some of which were known to be toxic and many of which were already considered adulterants. Indeed, the role of arsenic, a known toxin, in the manufacturing process of many of the new chemical dyes actually slowed down the recognition that synthetic dyes themselves might be toxic. Some chemists argued that the new synthetic dyes were a harmless and scientific way to color and preserve food, helping to increase the range and variety of food available to the wider public. Other chemists, however, claimed that the safety of the new dyes was not established, and that, like traditional colorings, they were being used illicitly to disguise the quality of food products and deceive the consumer.

These substances, initially viewed as "wonders of science," rapidly became associated with emerging public distrust of scientific authority. In these controversies, analysts contested among themselves, as well as with producers, retailers, and the public, what constituted legitimate or illicit food additives. The new synthetic dyes were never considered natural, unlike many of the other substances previously identified by chemists and the public as adulterants. More obviously than with other chemical products, their synthetic nature was thus yoked to scientific innovation.

As the mechanization and industrialization of food production developed apace in the later decades of the nineteenth century, new technologies and chemical additives presented new kinds of problems.[5] Chemists promoted chemistry as the main weapon in fighting food adulteration, and also helped to define what counted as adulteration. However, in their claim to offer rational and scientific ways of managing the increasing complexities of the food supply, they had to negotiate and compromise with producers, retailers, and politicians.[6]

At the same time, manufacturers, retailers, scientists, politicians, and social reformers all used the media, both editorial and advertising, to further their own interests. All types of periodicals played an important role in shaping the public's understanding of new discoveries in science, technology, and medicine. While news reporting to some degree reflected public opinion, how issues were reported in the media also helped form public perception, making the information flow a two-way process. This resulted in a far-from-passive appropriation of science by the public.[7]

The media set itself up as a mouthpiece for different constituencies of consumers and as its view of the dyes darkened so its calls for legislation increased. Chemists used the media as part of their efforts to capture positions of public authority and political prominence as advisors to the legislature. Several chemists, including dye manufacturer Ivan Levenstein, consultant William Crookes, and public analyst Charles Cassal, even set up their own publications. Meanwhile, manufacturers and retailers also turned to the media to promote the importance of industrial food production. Different constituencies vied with one another in trying to speak on behalf of the consumer.

Dye and food manufacturers responded defensively to the increasingly negative media coverage, by either denying knowledge of the use of the dyes in food production or arguing that the dyes were harmless, especially in the small quantities used. Companies also began to use scientists to validate their claims with BASF, for example, employing prominent chemists Hofmann, Virchow, and Cazeneuve to publicly state that many aniline and azo dyes were harmless.

The widespread use of undetectable dyes in food also posed a challenge for regulators. To ban all coloring of food would impose unacceptable restrictions upon consumer choice. The chemical dyes lay at the heart of a negotiation between freedom of trade, consumer choice, and the support of commercial practices on the one hand versus consumer protection, public health, and greater transparency on the other. However, assessing to what extent chemical additives should be used in food production depended on the commercial constraints of profit-making food producers and retailers, the opinions of chemists, and the political constraints imposed by the need to feed growing urban populations economically and safely. Food producers, consumers, health reformers, politicians, and the media all played a part in transforming how the dyes were understood and used beyond the chemical laboratories and factories.

While consumers initially welcomed science as an independent arbiter in the food adulteration debate, the public soon began to distrust its claims of neutrality.

The introduction of synthetic dyes, a "miracle" product of science, into food in the nineteenth century provides an interesting and instructive example of the type of public skepticism of science and scientists that is usually associated with the late twentieth century. For example, in a study of the Dutch consumer movement in the latter decades of the twentieth century, A. H. van Otterloo describes the increasing divide between the public, who view additives such as colorings, flavorings, and preservatives as adulteration, and scientists who see them as part of the legitimate food manufacturing process.[8] Exploring the introduction and mediation of the first chemical dyes used in food makes clear that the current ambiguity surrounding what is viewed as adulteration or innovation is not a recent phenomenon and that trust in scientists was never fully secured in the nineteenth century.

Testing problems inside and outside of the laboratory

The difficulty of reaching agreement around the application of new scientific objects in the commercial and public sphere intensifies as the number of vested interests grows, requiring a new approach to consensus building.[9] Individual chemists from different social and institutional settings battled to establish rules and platforms from which to know and normalize these novel chemical entities. Chemists across Europe and America had different views of the new dyes in food and drink, according to their social, political, economic, and institutional circumstances and status. The creation of separate societies and institutions within each country contributed to a growing distinction between different sectors of the chemical milieu, including academic, consulting, amateur, and retailing and wholesaling chemists and pharmacists.

The situation in assessing chemical dyes was particularly complex. These were novel and undetectable substances being created by chemists, themselves unable to agree on their nomenclature or chemical composition, and which were rapidly yet silently and invisibly proliferating throughout the international marketplace. This was a situation with uncertain facts and a widespread and disparate group of parties seeking agreement. The lack of consensus as to what constituted adulteration and the difficulty of knowing and identifying synthetic dyes and their toxicity meant that experts fell back on their authority and moral, political, and commercial priorities to settle disputes. While much historical work has examined how scientists asserted their authority in constructing facts and consensus in fields such as metrology and electricity, analytical chemistry arguably offers a more complex and interesting case study of experimenters' regress and the social negotiation of "scientific truths," although it has had far less attention.[10]

The difficulty for the analytical chemists was that by the 1880s there were hundreds of dyes in the European and American marketplace—some known, many unknown—and food and drink manufacturers often added several coloring additives to one product in order to make the detection of discrete dyes harder.[11] Testing for a known substance, or to eliminate a specific substance, is a much easier task than identifying the entire range of substances that might be present in any food or drink sample. Checking for the presence of several unspecified dyes would entail a whole battery of individual tests. This new testing regime proved to be a complex epistemological maneuver, with chemists mobilizing old repertoires of testing for a new context, cobbling together collections of tests from different specialties and chemical practices. Experimenters employed an array of analytical techniques to identify these new substances, from distillation, heating, filtration, acidic reagents, and titration to spectroscopy and microscopy, relying extensively on their senses of smell, taste, and sight. Chemists adapted and blended methodology and techniques from traditional analytical chemistry as well as existing industries and crafts, particularly textiles and dyemaking, together with new chemical and industrial techniques and emerging concepts of structural and synthetic chemistry.

The issues were just as complex for the toxicity experiments devised by physiological chemists using animal or in vitro testing. These tests involved using animals or test tubes as proxies for the human digestive system. Yet these types of modeling too were beset by a multitude of problems, including a large number of variables, nontransferability, and individual variation and tolerances of the consuming subject. Other problems included the difficulties of assessing long-term effects and deciding which dyes among hundreds to test for, and in what form. Tests involving the body and attempts to calibrate bodily senses such as taste and sight were particularly complex and compelling for nineteenth-century scientists. Individual dyes were used in food in minute quantities. Outside of the laboratory, very few of the substances were pure. In most cases, chemists were dealing with admixtures of chemicals, including known and unknown contaminants. Moreover, while the individual elements of the mixture might not be toxic, reactions between the dyes and other chemicals in the body might produce harmful effects. The fact that these chemicals were being consumed in many different unmonitored, and often unknown, ways with their properties and characteristics continually changing made their effects particularly hard to understand and predict.[12] While the increasing sophistication of epidemiology during this period enabled scientists to establish links between certain diseases and their causes, its use for assessing the risk of any number of small doses of chemicals in food or the environment over a long period of time was problematic, other than in occupational settings where variables and the presence of specific contaminants are more controlled. Chemists faced the problem of having to decide which type of dyes to target and in what context. Choosing a handful of dyes to represent the hundreds on the market was an almost impossible task. The variables, starting with those among the commercial dyes themselves, were manifold, which may partly explain the lack of consensus among scientists as to which dyes were to be considered harmful or not.

Those chemists charged with exposing risk were at a disadvantage compared with those producing the products, hindered by a lack of samples, a lack of knowledge about which dyes were being used, and secrecy surrounding their production and supply. Unlike the dyes themselves, increasingly manufactured in large state-of-the-art industrial research laboratories, experiments to detect their presence in food and their toxicity were conducted with far less funding and in a variety of settings, from institutional and educational facilities in university departments, to state-sponsored research facilities, to private laboratories, often located in chemists' homes and shops.[13]

The disorder of the real world made the promises of analytical chemists to clean up and rationalize the chemical world outside of the laboratory into hollow ones. The sheer numbers and diversity of the dyes overwhelmed the capacity of analysis to produce an ordered, named, legally visible body of material substances, able to be monitored and controlled.

Lack of standardization

Reaching consensus becomes more difficult the broader the decision-making group and social context become, and the chemists were just one part of the equation. Standardization becomes a far more complex situation to resolve within a competitive marketplace, where the negotiation includes manufacturers, retailers, the public, and politicians as well as scientists. Many problems arose from the multitude of different names, both chemical and commercial, used for the dyes. Reaching unanimity over both nomenclature and taxonomy involved a transformation in how chemists and others viewed these new substances and was never successfully achieved. The creators, manufacturers, wholesalers, retailers, users, and consumers of these dyes all conferred names upon them, leading to many synonyms being used for identical chemicals. Meanwhile, much reliance was placed on the visual and subjective observation of differing shades of color, yet the impossibility of agreeing upon a standardized color measurement framework suggests that grading color for identification purposes relied heavily on tacit and empirical skills of the chemist.[14]

The chemists were attempting to use experimental evidence for the purpose of classification in fields where experimental and theoretical consensus had not yet been secured. Experimental consensus surrounding the toxicity of the coal tar dyes similarly remained, and still remains, elusive. So experiment could be used neither for classification nor for proof. Chemists were operating within different disciplinary frameworks and within separate social and professional settings. When no agreement could be reached on establishing the facts of the case, including the experimental parameters and conditions necessary to obtain them, scientists fell back on issues of personal credibility, trying to secure support for their hypotheses and views through the media and the courts, and by lobbying peers, the public, and politicians. Social, economic, and political aspects played as large a part as the natural knowledge claims in a situation that was never securely resolved. Codification by way of standards is as much political as it is scientific. Countries, institutions, companies, and individuals looked to create standards for a variety of different reasons, from the control of trade to the establishment of authority.

What is also clear from researching the introduction of chemical dyes in food across different countries is that chemists acted as a crucial link between regulators and food producers and the public in determining the regulatory parameters surrounding the use of chemicals in food.[15] Chemists worked with the state and food companies to help legitimize the use of chemical dyes as food ingredients, changing the West's relationship with food.

Creating scientific expertise

Like the new substances themselves, science and scientists transformed within the public realm. And examining the unpredicted but widespread use of new chemical substances and how they were assessed and managed reveals how trust and "scientific" knowledge are formed and evaluated. On the one hand, chemists were discovering and manufacturing, on an industrial scale, new products that would transform consumables, while on the other hand they were being paid as public health advisers to police issues afflicting industrialized societies, such as food adulteration, environmental pollution, poor sanitation, and a contaminated water supply.[16]

The result is, as we have seen, an early attempt to create a system of trust surrounding industrialized food, at a time when certain ingredients were becoming legitimized in food production, a system that is still being challenged today.[17] Indeed, the legitimization of the industrial food supply, including the introduction of man-made substances such as chemical additives, is an early example of the creation of an "expert system."[18]

Harry Collins and Robert Evans claim, in their call for a "third wave" of science studies, that the demarcation between experts and other interested parties in society became increasingly blurred during the latter decades of the twentieth century. They argue that, because the political, public, and commercial demands for expertise usually run ahead of agreed-upon scientific knowledge, a renewed recognition and demarcation of different types of expertise was needed. However, the present book has shown that in the nineteenth century, as today, expertise, knowledge, and authority were contingent on prolonged and extensive debate and mediation between diverse groups, in different social, institutional, and geographical settings. Self-proclaimed experts such as public analysts attempted to draw boundaries between scientific and other expertise, whether legal, commercial, public, or political, but those boundaries were always blurred and open to negotiation, even among experts from within the same scientific discipline.

Chemists formed their expert opinions in differing political, commercial, and cultural situations, and public analysts were never neutral arbiters. Moreover, what counted as knowledge, and who claimed to have access to it, were issues as political and as open to negotiation in the late nineteenth century as Sheila Jasanoff and others have demonstrated they are in today's society.[19]

The scientists themselves were neither neutral nor disinterested. While

Weyl's interest in what impact the new dyes could have on the human body was not surprising considering his role as a hygienic and physiological chemist, the German chemist also recognized the benefits of testing the new chemical dyes in order to establish a consensus around their safe use. Such a program would legitimize the status of synthetic dyes, hygienic chemists, and Germany's industrial and scientific supremacy.[20] Similarly, the French chemist Cazeneuve was a politician as well as a consulting chemist who worked for a variety of businesses, including dye-manufacturing companies. Cazeneuve clearly saw the benefits of close cooperation between chemists and industry. Hygienic and physiological chemists identified toxicity testing of the new dyes as a way of expanding their repertoire and status, in a similar manner to the way in which public analysts had identified food adulteration as a means to improve their position and professional work earlier in the century. Analysts, meanwhile, continued to use food adulteration concerns to increase their own status. A prominent example was Wiley, who aligned himself with the American pure food movement to increase his bureaucratic empire and boost the importance of analytical chemists in the US.[21] Similarly, Lieber, the German-born, New York–based chemist, who advocated that every individual manufactured dye should be tested by its producer and sold as a food dye only if proved to be safe by exhaustive physiological tests, was aware of the benefits such a measure would have for companies such as his own.[22]

During this period, public analysts were very much at the center of disputes over authority, standards, and methodology in public health and were seldom disinterested. Chemists such as Hehner were paid to be consultants to businesses as well as public authorities and appeared as witnesses for both prosecution and defense in court. The capacity of consulting chemists to adjust their professional opinions to meet the criteria of their clients, whatever their personal views, is seen in the response and views of analysts in a variety of situations. Analysts expressed different views on the dyes in different settings, views that altered with their conflicting roles as public and political advisers, food custodians, scientific researchers, and commercial consultants. Chemists appeared as experts in courts, advisers to government committees, and commentators in the media in order to make a living and to ensure that science and scientists played an active role in social and political decision making.[23] Public analysts had to devise strategies to ensure that the uncertainties surrounding the dyes did not undermine their credibility as experts or the status of chemistry. Court strategies adopted by public analysts were part of their maneuver of falling back on socially constituted expertise when experiments fail. When the chemists were unable to resolve their arguments chemically, they turned to moral arguments of "purity" and "deception" to make their case. The courts became another arena, alongside the laboratory, parliament, marketplace, and home, in which the quality of food was determined and policed.

Chemists looked to tests and standards to create order and ensure that science, and scientists, could assert authority and integrity in a commercial world that threatened to exploit the plethora of novel dyes for financial gain at the possible risk to consumers and the public's confidence in these new scientifically created substances. However, expert systems of managing and controlling the changing environment of food production were the result of compromises between different communities and vested interests. At the same time, libertarian arguments surrounding consumer choice and free trade competed with debates surrounding public health and risk management.

Regulation and trade

Referring to nineteenth-century Britain, Sébastien Rioux argues that the state played a significant role in helping capitalist food production through legalizing certain forms of food adulteration.[24] The food journalist Joanna Blythman goes even further, contending that today's food manufacturers "rely on our regulators to fight their corner for them" when confronted with consumer criticism.[25] In the nineteenth-century cases explored in this book, food manufacturers turned to regulators to help support them in their industrialization of food, revealing how state regulation effectively legalized the use of chemical dyes as food ingredients while seeking to prevent their inappropriate use. By comparing the situation in different countries and cultures, however, it is possible to see how solutions for an increasingly global problem continued to be based on local factors, despite attempts to invoke universal natural laws to define and detect adulteration.

Regulators and public analysts were faced with difficulties associated with creating public policy about substances that were constantly changing and proliferating. One can ban and test for known substances but not for new, unknown ones, a situation that perplexed policymakers throughout the twentieth century and continues to do so. The lack of political consensus and the contestations surrounding the use of coal tar dyes in food arose partly because of the inability of chemists to generate reliable tests for detecting the presence of synthetic dyes in food and the fact that these new food ingredients proved so difficult to stabilize and standardize. Moreover, the chemists involved in arbitrating the use of chemical dyes in food differed in their own understanding of the new substances. An absence of agreement between chemists and their own allegiances to different sectors played a major part in ensuring that food regulations surrounding the use of chemical dyes in food and drink were never fully secured, and still vary from country to country today.

Cultural differences and a lack of consensus surrounding the new dyes, combined with the difficulty of regulating something already entrenched in the marketplace, thus led to a geographically mixed regulatory response. To some degree, legislation designed to limit and control the use of coal tar colors as food additives also reinforced their use.

The most precautionary response came from the US, which in 1907 specified just seven dyes to be used for food coloring. The main architect of the US laws was Harvey Wiley, a chemist who successfully harnessed consumers' and producers' concerns surrounding adulteration to secure antiadulteration legislation and empower the government's chemical department but still considered himself constrained by powerful vested interests.[26] Wiley and New York chemist Hesse worked closely with food producers and a handful of US dye manufacturers to produce a group of legally permitted dyes. This strategy of recommending a limited number of specified dyes for food use opened up a new legitimized and specialized market for food dyes, one eagerly embraced by US chemical companies as an opportunity to differentiate themselves from their bigger European dye-manufacturing competitors.

Germany and France took a different approach from the US, choosing to ban certain dyes known or considered to be toxic. This was a strategy encouraged by chemists, such as Weyl and Cazeneuve, who were anxious to reassure the market that the new dyes were a useful chemical product and that chemists could be relied upon to assist commerce in their proper use. This strategy led to the legitimization of coal tar colors as food dyes by creating an assumption that dyes not banned by law could be safely consumed. Indeed, wholesale chemists, particularly in Germany, the center of chemical dye production, exploited this situation by marketing dyes not named in the legislation as harmless. While municipal chemists complained about the increasingly widespread use of the dyes in food and drink and the uncertainty surrounding their safe consumption, other chemists, including chemists working for food and dye manufacturers, sought to reassure politicians and the public that most aniline and azo dyes were safe to use as food colorings. As a result, food regulation in France, Germany, and the US actually helped to establish and legitimize the presence of the new dyes in food production, creating opportunities for new markets and products and allowing food producers to successfully use regulation to their commercial advantage.

Britain, meanwhile, adopted an entirely different response in line with its nineteenth-century laissez-faire legacy, introducing no laws aimed specifically at coal tar dyes, despite growing media concern and two parliamentary inquiries into the matter. The lack of prescriptive legislation created a difficult situation for both public analysts and food manufacturers in Britain. In order to prevent the use of the dyes, public analysts had to prove in court that certain dyes were present in a product and either that they were being used to deceive the consumer or that their use could physically harm the consumer. As has been shown, tests both to detect the dyes and to prove their toxicity were inconclusive and easily challenged. At the same time, with no dyes considered safe to use, manufacturers chose not to disclose their use in food, making the analysts' position more difficult. Unlike some of the toxic metals previously used to color food, there was little evidence, and certainly no consensus, about the risks or benefits of using the new chemical dyes. The ability of any chemist to detect the new dyes in food, let alone establish their safety, was minimal. This was partly because analysts lacked both financial resources and standardized, robust analytical testing procedures to identify the hundreds of undisclosed dyes being used in food. Moreover, other scientists, including the government's excise chemists and chemists employed by food manufacturers, often disputed the accuracy of tests in court, challenging the authority and credibility of the public analysts.

A lack of status, authority, consensus, and technical expertise as well as a desire to endorse the reputation of chemistry, particularly the type of organic chemistry emanating from Germany at that time, probably all contributed to the reluctance of public analysts to pronounce judgment on the use of chemical dyes in food. Although public analysts had built their reputation initially on exposing the use of toxic mineral dyes in food, outright rejection of the synthetic dyes, being created by specialist organic chemists in well-equipped industrial laboratories, may have been a step too far. The chemical dyes were seen as a safer alternative than the mineral-based dyes formerly used to color food, initially grandstanded as examples of the power of chemistry to transform the world of consumption in this period. Synthetic dyes were entering the food supply system during a period when public analysts were fighting to secure their place as creditable and trustworthy experts as well as seeking to persuade the public that chemistry was the foremost sanitary and socially useful science. With no agreed-upon tests to identify or assess the safety of the new substances, it would not have been in the public analysts' interest or that of the reputation of chemistry itself to question the safety of science's latest invention. The new dyes were among a range of chemical additives that quickly became commodified and valuable ingredients in food. By the time concern began to be raised about coloring food and drink with chemical dyes, their use had become well established. This made the task of restricting them much harder because of the many vested interests concerned.

In each country examined, it is clear that different stakeholders looked to legislation to further their own interests. Food producers and retailers encouraged legislation that would effectively impose barriers on their competitors. Regulation also produced uniformity, which benefited large producers and retailers. Chemical manufacturers also benefited from legislation that effectively legitimized the use of certain coal tar dyes as food additives. Chemical analysts, meanwhile, used concern surrounding food adulteration and regulation to create an institutional base and to further their careers, reputation, and profile. Chemists, consumers, social reformers, free and fair trade advocates, manufacturers, and retailers all had a stake in forming economic and cultural knowledge and all invoked public health and consumer rights arguments to further their aims. While the legislation purported to represent natural knowledge, its actual formulation was strictly local and contingent rather than universal. Cultural and economic issues relating to free trade and protectionism played a much larger part in the legislation surrounding the chemical dyes than scientific claims.

Managing risk and reputation: the German nexus

More than any other nation, Germany had a vested interest in promoting chemical dyes as an important new resource in an increasingly industrialized international marketplace. By the end of the nineteenth century, Germany's chemical industry was the largest in the world and one of the most important industries and employers in Europe.

According to Ernst Homburg and Anthony Travis, it was in the 1950s that "historians began to identify the period 1850–1914 as one in which there was, for the first time, a highly productive convergence of science with technology, particularly in western Europe." The historical construction of what was termed the Second Industrial Revolution, much of it centered on the German chemical industry, was shaped by the requirements of the 1950s. This was a period defined by increased confidence in science, the Cold War struggle between capitalism and communism, and the aftermath of depression, with historians from J. D. Bernal to L. F. Haber focusing on the economic and social role of science.[27] During the latter decades of the twentieth century, historians reexamined the complex relationships between industry, technology, science, and society, recognizing that scientists are not detached and disinterested spokesmen, but have their own social agendas. Following dyes from their industrial production in Germany to their use in food across Europe and the US has shown us the roles, motivations, and relationships of chemists involved in industry, academic research, commerce, and public health.

Specifically, we have seen how professional and institutional communities play a significant role in legitimizing new science and mediating the risks created by the impact of industrialism on our material existence. While there may be substantial geographical distances between, say, communities of chemists and the different political and cultural contexts in which they operate, there still remain strong social and intellectual links in such expert systems. In the case of synthetic dyes, a clear German connection existed between many of the chemists operating in industrialized countries outside of Germany, encompassing academic and industrial chemists and public analysts. In Germany itself, a close network of chemists working in academia, the public sector, and industry was crucial to the movement of chemistry from the laboratory to the public sphere. Tracing the chemical dyes from their creation in German laboratories to their consumption as food coloring across four different countries has revealed an extensive chemical diaspora of German-born and German-educated chemists that helped to manage their journey and legitimize their transition from epistemic scientific objects to commercial objects of consumption.

Final words

Synthetic dyes, many of which were found during the twentieth century to be highly toxic, were never intended by their creators to be used as food additives. Their continued use for so long was the result of many factors, including a lack of consensus and status among scientists, political anxiety about food security, consumer expectations, and economic competition in the marketplace. Like many risks we face today, this unintended and unmonitored application of science occurred without any one group being in a position to grasp the full situation. Regulation and control became increasingly complicated as the practice was normalized and vested interest heightened and widened. Public analysts formed their expert opinions in a climate of diverse political, commercial, and cultural sentiment. What counted as knowledge, and who proclaimed to have access to it, was as political and fluid in the late nineteenth century as it is today.

Products and processes are transformed and used in countless ways by a multitude of actors once they leave the laboratory. Similarly, scientists themselves are not a homogeneous group, either inside or outside the laboratory, and are as much a product of their surroundings and circumstances as their creations become. Historians have commented on how scientific objects change from being epistemic objects invented by scientists from the moment they go out from the controlled environment of the laboratory into the wider world. This is clear with chemical dyes, which despite their bright colors became so hard to track down once they dissipated widely in the marketplace. While business, politicians, and the public all invoked science and chemists to represent their interests, the authority and technical ability of the scientists were never sufficient to successfully arbitrate on the issue, leaving the situation still unresolved more than a century later.

The beginnings of our modern risk management of food, employing scientists as independent advisers and adopting scientific methodology to assess food quality, became institutionalized in the late nineteenth century. However, the chemists appointed by the state were never totally impartial or independent, and effective scientific methodology, practices, and funding for monitoring and managing the risks was never secured. As a result, chemists, the state, manufacturers, retailers, and even consumers all helped to legitimize the introduction of synthetic dyes into our food, changing our relationship with food and paving the way for the legalization of thousands of new substances as food ingredients.

Acknowledgments

I would like to express my sincere thanks to a number of people without whom the publication of this book would not have been possible: to my husband, children, parents, siblings, and in-laws for their love and support; to my history of science mentors, Emma Spary and Hasok Chang, for their incredible knowledge, insight, and generosity over the years; to fellow foodie historian Lesley Steinitz for feeding, housing, and providing me with emotional and IT support during my research; to the educational institutions who fostered my intellectual curiosity over the years including Bath High and Brighton and Hove High Schools, Imperial College and University College London, and Darwin and Clare Hall at Cambridge University; to the many helpful librarians and archivists I have encountered; to editors Angela Creager, Karen Merikangas Darling, and Michael Koplow, and the anonymous reviewers for their comments, suggestions, and encouragement; and to Simon Schaffer and Sally Horrocks for recommending that I turn my research into a book.

Notes

INTRODUCTION

1. Layton and Pierce, *The Dawn of Technicolor*; Haines, *Technicolor Movies*.

2. Fricke, Scarfone, and Stillman, *The Wizard of Oz: The Official 50ᵗʰ Anniversary Pictorial History*.

3. Travis, *The Rainbow Makers*.

4. Blaszczyk, *The Color Revolution*; Blaszczyk and Spiekermann, *Bright Modernity*.

5. Baum, *American Fairytales*; Baum, *The Wonderful Wizard of Oz*; "Perkins's [*sic*] Purple."

6. Baum *The Art of Decorating*; Gaskill, "Learning to See with Milton Bradley"; Gaskill, "Vibrant Environments"; Simons Slater, "Little Geographies."

7. Rioux, "Capitalist Food Production"; White Junod, "The Chemogastric Revolution"; Kirchelle, "Toxic Confusion"; Stoff, *Gift in der Nahrung*; Von Schwerin, "Vom Gift im Essen zu chronischen Umweltgefahren."

8. In recent years, a handful of European historians have published research on the increasing use of toxic chemicals, including colorings and other ingredients, during the twentieth century. See particularly Stoff, *Gift in der Nahrung*, and Boudia and Jus, *Toxicants, Health, and Regulation since 1945*. The business historian Ai Hisano also has highlighted the introduction of synthetic dyes in her work on the US food industry's use of color in foods. See Hisano, *Visualizing Taste*.

9. Feingold, *Why Your Child Is Hyperactive*; Feingold, "Hyperkinesis and Learning Disabilities Linked to Artificial Food Flavors and Colors"; Stare, Whelan, and Sheridan, "Diet

and Hyperactivity"; Price et al., "In Vitro and in Vivo Indications of the Carcinogenicity and Toxicity of Food Dyes"; McCann et al., "Food Additives and Hyperactive Behavior."

10. Perkin, *Perkin Centenary London*; Garfield, *Mauve*; Travis, "Perkin's Mauve."

11. Pickering, "Decentering Sociology."

12. Blythman, *Swallow This*; Patel, *Stuffed and Starved*; Pollan, *The Omnivore's Dilemma*; Pollan, *Cooked*.

13. Coff, *The Taste for Ethics*.

14. Stanziani, "Negotiating Innovation"; Scholliers and Van den Eeckhout, "Hearing the Consumer?"; Teuteberg, *European Food History*; Burnett, *Plenty and Want*; French and Phillips, *Cheated Not Poisoned?*; Atkins, Lummel, and Oddy, *Food and the City*.

15. Leslie, *Synthetic Worlds*.

16. Houghton, *The Victorian Frame of Mind, 1830–1870*.

17. Rioux, "Capitalist Food Production."

18. Richards, "Legislation on Food Adulteration"; Richards, "Certain Provisions of Continental Legislation Concerning Food Adulteration"; "International Legislation on Adulteration of Food"; Teuteberg, "Adulteration of Food."

19. Brown, *Victorian News and Newspapers*; Rubery, *The Novelty of Newspapers, introduction*.

20. Altick, *The English Common Reader*; Beetham and Boardman, *Victorian Women's Magazines*.

21. Levenstein, suggesting that media involvement helped to inflame concerns and shape public responses, points out that it was the *Washington Post* that dubbed the chemical investigators appointed by Harvey Wiley, the bill's architect and promoter, the "Poison squad." Meanwhile, the *New York Times* commented that "few scientific experiments have attracted more attention from the general public than the chemical boarding house of the poison squad," where investigators consumed food adulterated with chemicals. Levenstein, *Fear of Food*; Anderson, *The Health of a Nation*; Young, *Pure Food*.

22. "Perkins's [*sic*] Purple," 222.

23. Holmes, "Beyond the Boundaries"; Fox and Nieto-Galan, "Natural Dyestuffs"; Fairlie, *Dyestuffs in the Eighteenth Century*; Nieto-Galan, *Colouring Textiles*.

24. Pickering, "Decentering Sociology."

25. Taussig, *What Color Is the Sacred?*

26. Klein, "Technoscience avant la Lettre."

27. Atkins, Lummel, and Oddy, *Food and the City*; Drouard and Oddy, *The Food Industries of Europe*; Shaw, "Changes in Consumer Demand and Food Supply"; Petrick, "The Industrialization of Food"; White Junod, "The Chemogastric Revolution."

28. Smith, *Pure Ketchup*; Fitzgerald, *Rowntree and the Marketing Revolution*; Corley, *Huntley and Palmers*.

29. Horrocks, "Consuming Science."

30. Hennessey, "Living in Color: The Potential Dangers of Artificial Dyes"; Schaeffer, "Color Me Healthy"; Spencer, "Choose Your Food by Colour—and Lose Weight!"

31. Hutchings, *Food Color and Appearance*, 5.

32. Ibid., 7.

33. Spence, "On the Psychological Impact of Food Colour."

34. For more on current debates about scientific expertise and the problems associated with using science and technology before any consensus in the scientific and technical community, see Collins and Evans, *Rethinking Expertise*.

35. Homburg, "The Rise of Analytical Chemistry"; Gee, "Amusement Chests and Portable Laboratories."

36. Pickstone, *Ways of Knowing*, 104.

37. Ibid.; Bud and Roberts, *Science Versus Practice*.

38. For more on reaching agreement with uncertain facts, see, for example, Collins, *Changing Order*; Gooding, Pinch, and Schaffer, *The Uses of Experiment*; Shapin and Schaffer, *Leviathan and the Air-Pump*; Pickering, *The Mangle of Practice*; Pickering, *Science as Practice and Culture*; Demortain, *Scientists and the Regulation of Risk*.

39. Spary, "Ways with Food," 763–71.

40. Stoff, *Gift in der Nahrung*.

41. Teuteberg, *European Food History*, 8; Spary, *Feeding France*; Merki, *Zucker gegen Saccharin*.

42. Mennell et al., *The Sociology of Food*; Lévi-Strauss, *Introduction to a Science of Mythology*; Bourdieu, *Distinction*; Bourdieu, Bruegel, and Atkins, "That Elusive Feature of Food Consumption"; Douglas, *Food in the Social Order*.

43. Goody, *Cooking, Cuisine and Class*; Mennell et al., *The Sociology of Food*; Mintz, *Tasting Food, Tasting Freedom*; Mintz, *Sweetness and Power*.

44. Ferrières, *Sacred Cow, Mad Cow*, preface, 231, 72.

45. Broomfield, "Rushing Dinner to the Table."

46. Wagner and Hassan, *Consuming Culture in the Long Nineteenth Century*.

47. See works such as Money, *The Destroying Angel*; Willis, *The Kellogg Imperative*; Schwartz, *Never Satisfied*.

48. Wessell, "Between Alimentary Products and the Art of Cooking."

49. Oddy, "Food Quality in London and the Rise of the Public Analyst."

50. For more on environmental and toxic risk, see Boudia and Jas, "Introduction"; Boudia, "Global Regulation"; Fressoz, "Gaz, Gazomètres, Expertises et Controverses"; Fressoz and Roux, "Protecting Industry and Commodifying the Environment."

51. Giddens, *The Consequences of Modernity*; Giddens, *Runaway World*; Adam, Beck, and Van Loon, *The Risk Society and Beyond*; Beck, *Risk Society*; Beck, "Critical Theory of World Risk Society."

52. Fressoz, "Beck Back in the 19th Century," for another comparison of nineteenth-century risk with Beck's description of twentieth-century risk.

53. Spiekermann, "Redefining Food."

54. Dessaux, "Chemical Expertise and Food Market Regulation"; Guillem-Llobat, "The Sugar Industry."

55. Phillips and French, "Adulteration and Food Law."

56. French and Phillips, "Sophisticates or Dupes?"

57. Hamlin, A Science of Impurity.

58. For an explanation of the historiographical debates surrounding different methodologies and disciplinary boundaries and an advocacy for their redundancy, see the authors' introduction in the 2011 reprint of Shapin and Schaffer, Leviathan and the Air-Pump. For an argument promoting more pluralism in the history and philosophy of science, see Chang, Is Water H$_2$O?

59. Bijker, Hughes, and Pinch, The Social Construction of Technological Systems.

60. See, for example, Jardine, Secord, and Spary, Cultures of Natural History; Golinski, Science as Public Culture; Hannaway, The Chemists and the Word; Holmes, "Beyond the Boundaries"; Nieto-Galan, Colouring Textiles; Roberts, "Filling the Space of Possibilities."

CHAPTER ONE

1. Clydesdale, "Color as a Factor in Food Choice"; Hoegg and Alba, "Taste Perception"; Hutchings, Food Color and Appearance; Keyser, Artificial Color; Lyman, A Psychology of Food; Spence, "On the Psychological Impact of Food Colour."

2. Wen, "Food Color Trumps Flavor."

3. Activated charcoal is used in filtration products or as a means of treating overdoses, as its holey surface allows it to absorb chemicals. Although used widely as a "detox" and an antibloating food ingredient, its medical properties in food are disputed. Medlin, "Activated Charcoal Doesn't Detox the Body"; Steen, "So, Eating 'Activated Charcoal' Is a Thing Now."

4. Public concerns about E numbers, especially artificial colors, increased from the mid-1980s with publications such as Hanssen, E for Additives.

5. The European Union E number scheme is based on the International Numbering System for food additives determined by the Codex Alimentarius, a collection of internationally recognized food standards. https://ec.europa.eu/food/safety/international_affairs/standard_setting_bodies/codex_en.

6. Blythman, Swallow This, 154–72; McCann et al., "Food Additives and Hyperactive Behaviour."

7. Filby, A History of Food Adulteration and Analysis.

8. Spary, Eating the Enlightenment; Spary, Feeding France.

9. Spary, Eating the Enlightenment; Spary, Feeding France.

10. Spary, Feeding France, 320.

11. Accum, A Treatise on Adulterations of Food and Culinary Poisons. A similar book published anonymously but believed to have been written by John Dingwall Williams was Deadly Adulteration and Slow Poisoning.

12. Sumner, "Retailing Scandal"; Burney, Poison, Detection, and the Victorian Imagination; Holloway, Royal Pharmaceutical Society of Great Britain.

13. Gee, "Amusement Chests and Portable Laboratories"; Homburg, "The Rise of Analytical Chemistry."

14. Homburg, "The Rise of Analytical Chemistry."

15. Chevallier, *Dictionnaire des Alterations et Falsifications*. Jean Baptiste Alphonse Chevallier was a French pharmacist and editor of the *Journal de Chimie Médicale, de Pharmacie et de Toxicologie*. Wisniak, "Jean Baptiste Alphonse Chevallier."

16. MacLeod, *Government and Expertise*; Rabier, *Fields of Expertise*.

17. For more on Liebig and Pasteur's work in industry and commerce, see Brock, *Justus von Liebig*; Debré, *Louis Pasteur*. For more on the expanding role of chemists in Britain and continental Europe, see Homburg, "The Rise of Analytical Chemistry"; Knight and Kragh, *The Making of the Chemist*; Bud and Roberts, *Science Versus Practice*; Russell, *Chemists by Profession*.

18. Charton, *Guide pour le choix d'un état, ou, Dictionnaire des profes-sions*, cited in Crosland, "The Organisation of Chemistry in Nineteenth-Century France," 11; Schödler, "Das chemische Laboratorium unserer Zeit," cited in Homburg, "Two Factions, One Profession," 39.

19. MacLeod, *Government and Expertise*; Rabier, *Fields of Expertise*; Hamlin, *A Science of Impurity*.

20. Douglas, *Purity and Danger*. For more on the sanitation movement in Britain, see Chadwick, "The British Sanitary Movement"; Hamlin, *Public Health*; Porter, *Health, Civilization, and the State*; Webster, *Victorian Public Health Legacy*.

21. Allen, *Cleansing the City*; Bashford, *Imperial Hygiene*; Kelley, *Soap and Water*; Hassall and Lancet Analytical Sanitary Commission, *Food and Its Adulterations*.

22. Carpenter, *Protein and Energy*; Spiekermann, "Twentieth-Century Product Innovations in the German Food Industry"; Spiekermann, "Redefining Food."

23. Filby, *A History of Food Adulteration and Analysis*, 217.

24. Whorton, *The Arsenic Century*, 153.

25. O'Shaughnessy, *Poisoned Confectionary*.

26. Gage, *Colour and Culture*; Nieto-Galan, *Colouring Textiles*; Ball, *Bright Earth*.

27. See Clayton, *Arthur Hill Hassall*; Hassall and Lancet Analytical Sanitary Commission, *Food and Its Adulterations*; Hassall, *Adulterations*.

28. Hassall and Lancet Analytical Sanitary Commission, *Food and Its Adulterations*, 4. See also page 177 for more detail.

29. Ibid., 4.

30. Ibid., xxx.

31. Ibid., xxxiv; Burney, *Bodies of Evidence*; Burney, *Poison, Detection, and the Victorian Imagination*; Whorton, *The Arsenic Century*.

32. Hassall and Lancet Analytical Sanitary Commission, *Food and Its Adulterations*, xxxv.

33. Whorton, *The Arsenic Century*, 155.

34. Dickens, *David Copperfield*, chap. 3.

35. Keene, *Science in Wonderland*.

36. Kingsley, *The Water-Babies*, 149. Henry Letheby was an analytical chemist and public health officer, a lecturer at the London Hospital, and a regular contributor to *The Lancet*.

37. Stern, "Adulterations Detected"; Stern, *Home Economics: Domestic Fraud in Victorian England*.

38. Stern, "Adulterations Detected."

39. Christopher Hamlin showed how effectively chemists such as Hassall and Edward Frankland used rhetorical skills and public propaganda to raise the status and authority of scientists by promoting the ability of science to address public health issues. Hamlin, *A Science of Impurity*.

40. On the geographical reorganization of food production and distribution in France in the twentieth century, for example, see Tenhoor, "Architecture and Biopolitics at Les Halles."

41. Ciritello, *Baking Powder Wars*.

42. Cobbold, "The Rise of Alternative Bread Leavening Technologies."

43. For more information about the introduction of synthetic sweeteners into food products, see De La Pena, *Empty Pleasures*, and Warner, *Sweet Stuff*.

44. Latour, *Reassembling the Social*; Shapin, *A Social History of Truth*; Deleuze and Guattari, *A Thousand Plateaus*; Delanda, *A New Philosophy of Society*.

45. Nordau, *Degeneration*; Walker, *Zola*, 89; Olson, *Science and Scientism*, 277–95.

46. Olson, *Science and Scientism*, 277–95.

47. Civitello, *Baking Powder Wars*; Meyer-Renschhausen and Wirz, "Dietetics, Health Reform, and Social Order"; Mintz, *Sweetness and Power*; Warner, *Sweet Stuff*; Mintz, "The Changing Roles of Food."

48. Kellogg, *The New Dietetics*.

49. Fletcher, *Fletcherism*.

50. Nissenbaum, *Sex, Diet, and Debility*; Griffith, "Apostles of Abstinence."

51. Whorton, *Crusaders for Fitness*.

52. Fischler, *L'homnivore*.

53. Beck, *Risk Society*; Setbon et al., "Risk Perception of the 'Mad Cow Disease' in France."

54. Beck, *Risk Society*; Ferrières, *Sacred Cow, Mad Cow*.

55. Atkins, "Sophistication Detected"; Scholliers, "Defining Food Risks and Food Anxieties throughout History"; French and Phillips, *Cheated Not Poisoned?*; Stanziani, "Negotiating Innovation"; Stanziani, "Information, Quality, and Legal Rules."

CHAPTER TWO

1. "Beautiful Tar: Song of an Enthusiastic," 123.

2. For more on alchemy, see Principe, *The Secrets of Alchemy*.

3. The alkaloid quinine is a naturally occurring chemical compound that was first isolated in 1832 from the bark of a cinchona tree, native to South America. Quinine had been used to treat malaria since the early seventeenth century.

4. For more about indigo, see Kumar, "Plantation Indigo and Synthetic Indigo"; Kumar, *Indigo Plantations and Science in Colonial India.*

5. Runge, "Ueber einige Produkte der Steinkohlendestillation."

6. Baeyer, *Ueber die chemische Synthese*; Campbell, *Chemical Industry.*

7. Homburg, "The Influence of Demand on the Emergence of the Dye Industry"; Nieto-Galan, *Colouring Textiles*, 182; Baeyer, *Ueber die chemische Synthese*; Hofmann, *On Mauve and Magenta*; Hofmann, *On the Importance of the Study of Chemistry.*

8. Travis, *The Rainbow Makers*; Travis, "From Manchester to Massachusetts via Mulhouse"; Travis, "Between Broken Root and Artificial Alizarin."

9. Ball, *Bright Earth.*

10. Nieto-Galan, *Colouring Textiles*, 192–93.

11. Kumar, "Plantation Indigo and Synthetic Indigo"; Kumar, *Indigo Plantations and Science in Colonial India.*

12. Kumar, "Plantation Indigo and Synthetic Indigo"; Kumar, *Indigo Plantations and Science in Colonial India*, 186; Homburg, "The Influence of Demand on the Emergence of the Dye Industry"; Travis, "Between Broken Root and Artificial Alizarin"; Travis and Brent Schools and Industry Project, *The Colour Chemists.*

13. Homburg, "The Influence of Demand on the Emergence of the Dye Industry."

14. Garfield, *Mauve*, 43–45.

15. Ibid., 65–67.

16. Blaszczyk, *The Color Revolution*, 22–26.

17. "Perkins's [*sic*] Purple."

18. Ibid.

19. Halliday, *The Great Stink of London*; Hamlin, *A Science of Impurity.*

20. For an example of the type of debate in the press, see "The Thames and Its Deodorization," 569, in which Henry Letheby, a chemist and trained doctor who worked as an analytical chemist and public health officer, is reported as disputing the attempts to "deodorise" the Thames by Hofmann and Edward Frankland, then chemistry lecturer at St. Bartholomew's hospital. Letheby claimed that the agent his peers were planning to use could be contaminated with arsenic and could cause more damage. The article refers to the inability of the scientists to produce consistent evidence and analysis.

21. Fitzgerald, *The Hundred-Year Lie*; White, "Chemistry and Controversy."

22. Beer, *Open Fields*; Sheets-Pyenson, "Popular Science Periodicals in Paris and London."

23. Orland, "The Chemistry of Everyday Life."

24. Fox, *Dye-Makers of Great Britain*, chap. 4.

25. Fyfe and Lightman, *Science in the Marketplace.*

26. Lightman, *Victorian Science in Context*, 2; Paradis, *Victorian Science and Victorian Values.*

27. Cantor et al., *Science in the Nineteenth-Century Periodical*, preface.

28. Lightman, *Victorian Science in Context*, 187.

29. For more on Hofmann's work and role as an advocate for science, education, and industry, see Meinel and Scholz, *Die Allianz von Wissenschaft und Industrie*.

30. Hofmann, *On the Importance of the Study of Chemistry*.

31. Hofmann, *On Mauve and Magenta*.

32. "Scientific Facts," 116.

33. For more about a nineteenth-century female audience for sciences, see Shteir, *Cultivating Women, Cultivating Science*.

34. Orland, "The Chemistry of Everyday Life."

35. Findling and Pelle, *Encyclopedia of World's Fairs and Expositions*; Teughels and Scholliers, *A Taste of Progress: Food at International and World Exhibitions*; *International Exhibition, 1862: Reports by the Juries* (1863).

36. "A Ramble into the Eastern Annexe of the International Exhibition," 342. As historians such as Martin Rudwick have shown, the far-distant geological history of the world held a similar fascination for many Victorians in the middle of the nineteenth century. Rudwick, *Earth's Deep History*.

37. "A Ramble into the Eastern Annexe of the International Exhibition," 342.

38. For more about the early reception and use of synthetic dyes in Victorian fashion, see Nicklas, "Splendid Hues"; Nicklas, "New Words and Fanciful Names."

39. Kortsch, *Dress Culture in Late Victorian Women's Fiction*; Wilson, *Adorned in Dreams*; Gernsheim, *Victorian and Edwardian Fashion*.

40. Gernsheim, *Victorian and Edwardian Fashion*, 53.

41. Fox, *Dye-Makers of Great Britain 1856–1976*, 96.

42. Taine, *Taine's Notes on England*, 19, 20.

43. Brock, *The Case of the Poisonous Socks*, chap. 1.

44. "Mythology and Socks," 160.

45. Noakes, "*Punch* and Comic Journalism in Mid-Victorian Britain."

46. "Accidents and Offences," 402.

47. "The Poisoned Hat," 262.

48. "A Hat That Was Felt," 90. *Funny Folks* was a weekly comic newspaper with four pages of cartoons and four pages of text. Designed for an adult readership, it described itself as "The Comic Companion to the Newspaper." See Hunt, *International Companion Encyclopedia of Children's Literature*.

49. Rubery, *The Novelty of Newspapers*.

50. "The Useful Book," *The Ladies' Treasury*, January 1, 1875, 42.

51. See Travis, *The Rainbow Makers*, 124, for a description of the arsenic method for aniline dye production.

52. "The Letter Box."

53. "Talk with Our Readers," 272.

54. Burney, *Poison, Detection, and the Victorian Imagination*; Whorton, *The Arsenic Century*.

55. Whorton, *The Arsenic Century*, 359.

56. Travis, "Poisoned Groundwater and Contaminated Soil."

57. "Poisoning by Coloured Silk Stockings," *The Englishwoman's Domestic Magazine* 164, August 1, 1878, 109.

58. "By All That's Blue," 38. Dyes discovered in the 1870s include phlox-ine, known as rose Bengal (discovered by Nolting, 1875); alizarin orange (Strobel and Caro, 1876); anthracene brown (Seubelich, 1877); various or-anges (Roussin and Poirrier, 1877–78); Biebrich scarlet (Nietzki, 1878); cloth reds (Ohler, 1879); Ponceau S (Pfaff and Nietzki, 1880). Ball, *Bright Earth*; Brunello, *The Art of Dyeing in the History of Mankind*.

59. "Colour and Design in Ornamental Needlework," 43.

60. Cantor et al., *Science in the Nineteenth-Century Periodical*, 18–19.

61. Murmann, *Knowledge and Competitive Advantage*, 25, 37.

62. Beer, *The Emergence of the German Dye Industry*.

63. Redlich, *Die volkswirtschaftliche Bedeutung der deutschen Teerfarben-industrie*, 18; Murmann, *Knowledge and Competitive Advantage*, 33.

64. "Queer Street," 79.

65. Travis, *The Rainbow Makers*; Beer, *The Emergence of the German Dye Industry*; Paul, *From Knowledge to* Power; Smith, *The Emergence of Modern Business Enterprise in France*.

66. Kumar, *Indigo Plantations and Science in Colonial India*.

67. Nieto-Galan, *Colouring Textiles*, 18.

68. Rocke, *The Quiet Revolution*.

69. Brock, *Justus von Liebig*; Hofmann, *The Life-Work of Liebig*; Holmes, "The Complementarity of Teaching and Research in Liebig's Laboratory"; Liebig, *Researches on the Chemistry of Food*; Meinel and Scholz, *Die Allianz von Wissenschaft und Industrie*; Haber, *The Chemical Industry During the Nineteenth Century*.

70. Roberts, "Bridging the Gap between Science and Practice."

71. Haber, *The Chemical Industry During the Nineteenth Century*; Meinel and Scholz, *Die Allianz von Wissenschaft und Industrie*.

72. Haber, *Chemical Industry, 1900–30*; Ben-David and Ben-David, *Centers of Learning*; Fox and Guagnini, *Education, Technology and Industrial Performance in Europe, 1850–1939*.

73. Berghahn, *Imperial Germany, 1871–1918*; Vaupel, "Wissenschaft und Patriotismus."

74. Ash and Surman, *The Nationalization of Scientific Knowledge in the Habsburg Empire, 1848–1918*; Phillips, *Acolytes of Nature*; Rocke, *The Quiet Revolution*; Vaupel, "Wissenschaft und Patriotismus."

75. Hugill and Bachmann, "The Route to the Techno-Industrial World Economy and the Transfer of German Organic Chemistry to America," 162.

76. Phillips, *Acolytes of Nature*; Vaupel, "Napoleons Kontinentalsperre und ihre Folgen."

77. Henderson, *The State and the Industrial Revolution in Prussia 1740–1870*.

78. Turner, "The Growth of Professorial Research in Prussia"; Ben-David, *The Scientist's Role in Society*.

79. Lenoir, "Revolution from Above"; Tuchman, *Science, Medicine, and the State in Germany*; Turner, "The Great Transition and the Social Patterns of German Science."

80. Rocke, *The Quiet Revolution*; Brock, "Breeding Chemists."

81. Schröter and Travis, "An Issue of Different Mentalities."

82. Martius, *Chemische Erinnerungen aus der Berliner Vergangenheit: zwei akademische Vorträge*, cited in Murmann, *Knowledge and Competitive Advantage*, 75.

83. Meinel and Scholz, *Die Allianz von Wissenschaft und Industrie*.

84. For a detailed explanation of the importance of patent law to the German synthetic dye industry, see Murmann, *Knowledge and Competitive Advantage*.

85. Davis, *Corporate Alchemists*, 60–63.

86. Murmann, *Knowledge and Competitive Advantage*. For more on the rise of theoretical organic chemistry and its application in industrial chemistry as viewed in the 1950s, see Taylor, *A History of Industrial Chemistry*, chaps. 15 and 16.

87. Belt and Rip, "The Nelson-Winter-Dosi Model and Synthetic Dye Chemistry," 129–54.

88. Caro, *Über die Entwicklung der Teerfarben-Industrie*.

89. Murmann, *Knowledge and Competitive Advantage*, 70–71.

90. Schoonhoven and Romanelli, *The Entrepreneurship Dynamic*, 184; Simon, "The Swiss Chemical Industry," 17.

91. Murmann, *Knowledge and Competitive Advantage*.

92. Ibid., 85.

93. Krätz, "Der Chemiker in den Gründerjahren," 269–70; Cocks and Jarausch, *German Professions, 1800–1950*, 13; Rüschemeyer, "Professional Autonomy and the Social Control of Expertise," 38–58; McClelland, *The German Experience of Professionalization*.

94. Johnson, "Academic Self-Regulation and the Chemical Profession in Imperial Germany," 241–71; Cocks and Jarausch, *German Professions, 1800–1950*; McClelland, *The German Experience of Professionalization*; Burchardt, *Professionalisierung oder Berufskonstruktion?*

95. Schröter and Travis, "An Issue of Different Mentalities."

96. Engel, *Farben der Globalisierung*; Engel, "Colouring Markets"; Engel, "Colouring the World"; Murmann, *Knowledge and Competitive Advantage*; Haber, *The Chemical Industry During the Nineteenth Century*; Beer, *The Emergence of the German Dye Industry*; Homburg, Travis, and Schröter, *The Chemical Industry in Europe*; Travis and Brent Schools and Industry Project, *The Colour Chemists*; Fox, *Dye-Makers of Great Britain 1856–1976*.

CHAPTER THREE

1. Engel, *Farben der Globalisierung*; Engel, "Colouring Markets"; Blaszczyk and Spiekermann, *Bright Modernity*.

2. Detector, "Letter to the Editor."

3. "Advertisement," *The Leeds Mercury*, September 27, 1865.

4. Jutta Kissener, corporate history, BASF emails to Carolyn Cobbold, May 8, 2012, and July 11, 2015.

5. Bernhard C. Hesse, "Letter to Frederick Dunlap" (Washington DC, August 7, 1907), General Correspondence, Bureau of Chemistry, Record Group

97, US National Archives, cited in Hochheiser, "Synthetic Food Colors in the United States."

6. Bernhard C. Hesse, "A Letter to Frederick Dunlap" (Washington DC, August 8, 1907), General Correspondence, Bureau of Chemistry, Record Group 97, US National Archives, cited in Hochheiser, "Synthetic Food Colors in the United States."

7. Hochheiser, "Synthetic Food Colors in the United States."

8. For more on cookbooks of the period, see Humble, *Culinary Pleasures*.

9. "Some Very Ancient Things," 242. For more about synthetic flavorings, see the work of US historian Nadia Berenstein: *Flavor Added* (blog), www.nadiaberenstein.com/blog; Berenstein, *The Inexorable Rise of Synthetic Flavor*.

10. "Aniline Colours," 3.

11. Ibid.

12. "Poisoned Candies," *The Health Reformer* 6–7, 1871, 131. The only other reference to a substance known as amboline in this period that I can find is a hair dye manufactured by two New York chemists from the mid-1860s and marketed as Kendall's Amboline. Aniline and azo dyes were increasingly used as hair dyes during this period. Fyke, *The Bottle Book*.

13. Hassall, *Food*, 258.

14. "Poisonous Ice Cream," 311.

15. For more on the coloring of oleomargarine, see chaps. 6 and 8.

16. "Meat-Tints for the Million," 101.

17. *The Country Gentleman: Sporting Gazette and Agricultural Journal* 1076 (1882): 1322; "Talk about Depression in Trade."

18. "Chit-Chat," 5.

19. "Blood Oranges," 309.

20. "How We Are Poisoned in 1890," 92.

21. "Advertisement for Eno's."

22. Albala, keynote address; Humble, *Culinary Pleasures*.

23. Broks, *Media Science Before the Great War*; Young, *Darwin's Metaphor*; Cantor et al., *Science in the Nineteenth-Century Periodical*, 24.

24. Silver, "Virchow, the Heroic Model in Medicine."

25. Cazeneuve, *Les Colorants de La Houille*; Paul, *From Knowledge to Power*, chap. 5.

26. Hesse, *Coal-Tar Colors Used in Food Products*, 35, 76, 88.

27. "BR Brückner Advertisement. Collection of Advertisements." (Berlin), Imperial Health Office R86/2255, Deutsche Archive. ("Daß ich, um das Ergrauen der aus rohem Fleische hergestellten Wurstwaaren zu verhindern, jedes zu dieser Wurst gehörige kilogramm Salz mit 20 gramm der von BR Brückner in Steglitz hergestellten conc. Wursttinctur oder Pulver (Carminsurrogat) untermenge. Diese surrogate entspricht ¶1 der reichsgesetzlichen Bestimmungen vom 5 Juli 1887 und ist der Gesundheit nicht nachtheilig.")

28. Hochheiser, "Synthetic Food Colors in the United States"; H. Kohnstamm, "The Development of Certified Pure Food Colors"; "H. Kohnstamm & Co."; Powers, "The Early Industrial Achievements of the Schoelkopf Family."

29. Hochheiser, "Synthetic Food Colors in the United States," 23–24; Hesse, *Coal-Tar Colors Used in Food Products*, 55, 178.

30. "60 Years of Dyestuff Manufacture by Williams (Hounslow) Ltd., Hounslow Middlesex" (London, 1936), Williams Bros Archives, Hounslow Library; Rideal, "An Investigation of Certain Substances Used in Colouring Foods."

31. "60 Years of Dyestuff Manufacture by Williams (Hounslow) Ltd., Hounslow Middlesex."

32. Ibid.

33. Evidence of Robert McCracken, managing director of United Creameries Ltd. 20 Dec 1899. *Report of the Departmental Committee Appointed to Inquire into the Use of Preservatives and Colouring Matters in the Preservation and Colouring of Food: Together with Minutes of Evidence, Appendices and Index* (London: H.M.S.O., 1901), 99–103.

34. Ibid.

35. For more on Keiller, see Atkins, "Vinegar and Sugar."

36. Testimony of L. K. Boseley, analyst to Keiller and Son. Ltd., 17 November, 1899, Report of the Departmental Committee Appointed to Inquire into the Use of Preservatives and Colouring Matters in the Preservation and Colouring of Food (London: H.M.S.O., 1901), 30–37; Weber, *American Chemical Journal* 18 (1896): 1092–96.

37. Testimony of L. K. Boseley, 30–37.

38. Drummond and Wilbraham, *The Englishman's Food*; Atkins, "Vinegar and Sugar."

39. Testimony of L. K. Boseley, 30–37.

40. French and Phillips, *Cheated Not Poisoned?*; Phillips and French, "Adulteration and Food Law"; Atkins, Lummel, and Oddy, *Food and the City*.

41. *Food and Sanitation, Formerly Food, Drugs and Drink*, December 16, 1893, 387.

42. "Forms of Adulteration," 5.

43. Ibid.

44. Hehner, *Analyst* 15 (1890): 221–26.

45. "Forms of Adulteration," 5.

46. Ibid.

47. For more about regulation in Germany, see Hierholzer, "Searching for the Best Standard." For more on regulation in France, see Stanziani, "Information, Quality, and Legal Rules"; Stanziani, "La Mesure de la Qualité du Vin en France, 1871–1914"; Dessaux, "Chemical Expertise and Food Market Regulation in Belle-Epoque France." For a comparison of Europe and the US, see Spiekermann, "Redefining Food." For Britain, see French and Phillips, *Cheated Not Poisoned?*; Atkins, *Liquid Materialities*; Miller, "Food Colours."

48. Collins and Oddy, "The Centenary of the British Food Journal"; Oddy, "Food Quality in London and the Rise of the Public Analyst, 1870–1939," 95–96.

49. "Editorial," *Food, Drugs and Drink* 1, 24 (1893): 3.

50. "Editorial," *Food, Drugs and Drink* 2, 27 (1893): 1.

51. Ibid.

52. Atkins, Lummel, and Oddy, *Food and the City*; Barker, Crawford Mackenzie, and Yudkin, *Our Changing Fare*; Boswell, *JS 100*; Corley, *Huntley and Palmers*; Heer, *Nestle*; Petrick, "'Purity as Life"; Zeide, *Canned*.

53. Civitello, *Baking Powder Wars*; Warner, *Sweet Stuff*.

54. Warner, *Sweet Stuff*.

55. For more on the structural changes in food production and distribution in Victorian Britain, see Fraser, *The Coming of the Mass Market, 1850–1914*; Winstanley, *The Shopkeeper's World 1830–1914*; Jefferys, *Retail Trading in Britain, 1850–1950*; Collins, "Food Adulteration and Food Safety in Britain in the 19th and Early 20th Centuries"; Shaw, "Changes in Consumer Demand and Food Supply."

56. Atkins, Lummel, and Oddy, *Food and the City*; Grüne, *Anfänge staatlicher Lebensmittelüberwachung in Deutschland*; Hierholzer, *Nahrung nach Norm*; Horrocks, "Consuming Science."

57. Rowntree Archives.

58. H. I. Rowntree, "Letter to A. H. Hassall" (York, October 2, 1888), H. I. Rowntree & Co. The Cocoa Works 1887–1904, HIR/1/15, Rowntree Archives, Borthwick Institute. An almost identical letter was sent on October 27, 1888, to J. F. W. Hodges in Belfast.

59. H.I. Rowntree, "Letter to Messrs Hassall & Clayton" (York, October 16, 1888), H.I. Rowntree & Co. The Cocoa Works 1887–1904. HIR/1/15, Rowntree Archives, Borthwick Institute.

60. Grivetti and Shapiro, *Chocolate*, 190.

61. Rowntree Archives.

62. Ibid.

63. H. I. Rowntree, "Letters to Ehrenfest & Co. Blackfriars" (York, February 1892), H. I. Rowntree & Co. The Cocoa Works 1887–1904 HIR/1/15, Rowntree Archives, Borthwick Institute.

64. Ehrenfest & Co., "Memorandum to H. I. Rowntree" (York, June 21, 1893), H. I. Rowntree & Co. The Cocoa Works 1887–1904, HIR/1/15, Rowntree Archives, Borthwick Institute.

65. "Rowntree Recipe Books" (York, various dates), H. I. Rowntree & Co. The Cocoa Works 1887–1904, HIR/7B/2, Rowntree Archives, Borthwick Institute.

66. British Government, *Report of the Departmental Committee Appointed to Inquire into the Use of Preservatives and Colouring Matters in the Preservation and Colouring of Food*, 61.

67. Ibid., paragraphs 60 and 61.

68. Atkins, *Liquid Materialities*, 181.

69. J. W. Rowntree, "Private Experimental Notebooks" (York, various dates), H. I. Rowntree & Co. The Cocoa Works 1887–1904, HIR/7b/10, Rowntree Archives, Borthwick Institute.

70. Alfred Ashby, "Letter to Walter Palmer Regarding Analysis of Colouring Matter in Biscuits" (Reading, 1891), HP/143, Huntley and Palmer Archive, Museum of English Rural Life.

71. S. Allen, "Notes on the History of the Chemical Department" (York, 1947), R/DT/CC/6, Rowntree Archives, Borthwick Institute.

72. Ibid.

CHAPTER FOUR

1. For more about contemporary debates about risk and regulation and the role of scientists, including in the food industry, see: Demortain, *Scientists and the Regulation of Risk*; Adam, Beck, and Loon, *The Risk Society and Beyond*; Beck, *Risk Society*; Ferrari, *Risk Perception, Culture, and Legal Change*; Hutter, *Anticipating Risks and Organising Risk Regulation*; Hutter, *Managing Food Safety and Hygiene*.

2. Weyl, *Handbuch der Hygiene*, 385.

3. Arata, "Detection of Colouring Matters in Wine"; Belar, "Detection of Foreign Colouring Matters in Red Wines"; Geisler, "A Delicate Test for the Detection of a Yellow Azo Dye Used for the Artificial Coloring of Fats"; Halpen, "The Detection of Foreign Colouring Matters in Preserved Tomatoes"; Micko, "Artificial Colouring of Oranges"; Rinzand, "Artificial Colouration of Wine"; Rota, "A Method of Analyzing Natural and Artificial Organic Colouring Matters"; Spaeth, "On Fruit Juices and Their Examination"; Spaeth, "The Detection of Artificial Colouring Matters in Sausages."

4. The variety of overlapping techniques and methodology employed suggest this was a period of transition between chemists' use of senses and a more "modern" reliance on quantitative laboratory results, a distinction that has been articulated by the historian Lissa Roberts, see Roberts, "The Death of the Sensuous Chemist." For more information about the knowledge and practices of analytical chemistry during this period, see Szabadváry, *History of Analytical Chemistry*.

5. Fagin, *Tom's River*.

6. Fitzgerald, *The Hundred-Year Lie*; Carson, *Silent Spring*.

7. One of the areas of taxonomy most discussed in the history of science is that of biological taxonomy. Carl Linnaeus's 1775 publication *Systema Naturae* introduced a detailed and universally adopted system of classification that remained largely unquestioned and intact until recent developments in DNA and evolutionary theory. Linnaeus, Stearn, and Heller, *Species Plantarum: Carl Linnaeus*; Anderson, *Carl Linnaeus*.

8. It was not until the 1850s when the newly synthesized coal tar dyes began to be widely used in industry and commerce, attracting many different ways of knowing and naming them. For a European-wide historical examination on the use of language and nomenclature in chemistry, see Bensaude-Vincent and Abbri, *Lavoisier in European Context*.

9. Hepler-Smith, "Just as the Structural Formula Does."

10. Ibid.

11. Muter, "On the Processes and Standards in Use at the Municipal Laboratory of the City of Paris."

12. Hesse, "The Industry of the Coal-Tar Dyes"; Redlich, "Die volkswirtschaftliche Bedeutung der deutschen Teerfarbenindustrie"; Murmann, *Knowledge and Competitive Advantage*, 248.

13. Nicklas, "New Words and Fanciful Names."

14. Jones, *German Colour Terms*; Schaeffer, *Die Entwicklung der künstlichen organischen Farbstoffe.*

15. For more on Linnaean classification, see Blunt, *Linnaeus*; Linnaeus, Stearn, and Heller, *Species Plantarum: Carl Linnaeus*; Koerner, *Linnaeus*; Anderson, *Carl Linnaeus.*

16. Jones, *German Colour Terms*, 132.

17. Weyl and Leffman, *The Coal-Tar Colors*, 17, 89–90.

18. Hueppe, *Die Methoden der Bakterien-Forschung*, 105.

19. Erdmann, Review of Schultz and Julius, *Tabellarische Übersicht der Künstlichen organischen Farbstoffe*, 767; Green, Schultz, and Julius, *A Systematic Survey of the Organic Colouring Matters.*

20. Rocke, *Nationalizing Science*; Klein, *Experiments, Models, Paper Tools*; Klein and Reinhardt, *Objects of Chemical Inquiry.*

21. Murmann, "Knowledge and Competitive Advantage"; Abelshauser et al., *German Industry and Global Enterprise*; Homburg, Travis, and Schröter, *The Chemical Industry in Europe.*

22. In the preface to his second edition in 1904, Green noted that since the first edition, published in 1894, "many of the older dyestuffs and older methods of manufacture have become obsolete, whilst newer processes have been introduced, new intermediate products discovered, and an enormous array of new colouring-matters have been brought into commerce."

23. Green, Schultz, and Julius, *A Systematic Survey of the Organic Colouring Matters*, viii.

24. The classification of chemicals remained a contentious issue throughout the nineteenth century, with chemists choosing many different systems to group dyes. Several other classification books based on Schultz's original tables were published in English and German using different methods of classification from Green. For more information, see Clark, *Handbook of Textile and Industrial Dyeing.*

25. Burdett, "The Colour Index."

26. Clark, *Handbook of Textile and Industrial Dyeing*; Burdett, "The Colour Index."

27. For examples of the various color measurement systems currently used in the food industry, see Good, "Methods of Measuring Food Color"; Caivano and Buera, *Color in Food.*

28. For more on the Munsell color system, see Cleland, *A Practical Description of the Munsell Color System*; Kuehni, "The Early Development of the Munsell System"; Munsell, *A Color Notation*; Munsell, "A Pigment Color System and Notation"; Nickerson, "History of the Munsell Color System, Company, and Foundation"; Landa and Fairchild, "Charting Color from the Eye of the Beholder."

29. Gaskill, "Learning to See with Milton Bradley"; Blaszczyk, *The Color Revolution.*

30. Gibson, *The Lovibond Color System*; Lovibond, "On the Scientific Measurement of Colour in Beer"; Johnston, *A History of Light and Colour Measurement*; Bud and Warner, *Instruments of Science.*

31. Gibson, *The Lovibond Color System*.

32. Rossi, "Let's Go Color Shopping with Charles Sanders Peirce."

33. Johnston, *A History of Light and Colour Measurement*; Cochrane, *Measures for Progress*. As explained in other chapters, the color of butter and oleomargarine was a contentious issue, with the dairy industry frequently complaining that the coloring of margarine yellow, to make it resemble butter, was an act of adulteration and fraud.

34. Johnston, *A History of Light and Colour Measurement*.

35. Ibid.

36. Cochrane, *Measures for Progress*, 270.

37. Mari, "Epistemology of Measurement."

38. Tal, "Old and New Problems in Philosophy of Measurement"; Chang, "Circularity and Reliability in Measurement"; Chang, *Inventing Temperature*; Cartwright, *Nature's Capacities and Their Measurements*; Chang and Cartwright, "Measurement"; Gooday, *The Morals of* Measurement; Schaffer, "Metrology, Metrication, and Victorian Values"; Schaffer, "Late Victorian Metrology."

39. For more on the development of analytical laboratories and the rise of the analytical chemist, see Gee, "Amusement Chests and Portable Laboratories." For more on chemical analysis in different fields and its contestation in the courts and other spheres, see Brimblecombe, *The Big Smoke*; Gerber, *Chemistry and Crime*; Hamlin, *A Science of Impurity*; Burns, "Analytical Chemistry and the Law."

40. For more on titration tests for industrial chemicals, see Szabadváry, *History of Analytical Chemistry*. For evidence on the use of indicator solutions to detect content from color changes in the sixteenth century, see Boas, *Robert Boyle and Seventeenth-Century Chemistry*; Debus, "Solution Analyses Prior to Robert Boyle"; Debus, "Sir Thomas Browne and the Study of Colour Indicators."

41. Stanziani, "Information, Quality, and Legal Rules"; Stanziani, "Negotiating Innovation"; Stanziani, "La Mesure de la Qualité du Vin en France, 1871–1914."

42. Gautier, "The Fraudulent Colouration of Wines"; Gautier, "Continuation of The Fraudulent Colouration of Wines." Gautier's articles were based on his work previously published in *Bulletin de la Société Chimique de Paris*.

43. Fell and Rocke, "The Chemical Society of France in Its Formative Years, 1857–1914." Armand Gautier (1837–1920) studied chemistry in Montpellier under J. A. Béchamp before moving to Paris, where he was a student of Adolphe Wurtz. In 1874, Gautier became director of the new medical laboratory at the Faculté de Médecine before succeeding Wurtz as professor of medical chemistry in 1884. His work included the isolation of isonitriles (isomers of nitrites), which he called carbylamines, in a double decomposition reaction between silver cyanide and simple or compound ether in 1866. He also developed methods for detecting and quantifying small amounts of arsenic and demonstrated that arsenic traces could be found in healthy animals. Thorpe, "Obituary of Gautier." For more on Wurtz and his school of structural chemists, see Rocke, *Nationalizing Science*.

44. Gautier, "The Fraudulent Colouration of Wines," 109–12.

45. Gautier, "Continuation of The Fraudulent Coloration of Wines," 130–35.

46. For more on food adulteration test centers in France, see Stanziani, "Information, Quality, and Legal Rules"; Stanziani, "Negotiating Innovation"; Stanziani, "La Mesure de la Qualité du Vin en France, 1871–1914"; Atkins, Lummel, and Oddy, *Food and the City*; Scholliers and Van den Eeckhout, "Hearing the Consumer?"; Scholliers, "Defining Food Risks and Food Anxieties throughout History." For more on the Municipal Laboratory in Paris, see chap. 7.

47. Muter, "On the Processes and Standards in Use at the Municipal Laboratory of the City of Paris." Muter was a founder of the *Analyst* and its coeditor from 1877 to 1891. He was public analyst for several boroughs in south London as well as for the county of Lincolnshire and principal of the South London School of Pharmacy. Obituary, *Journal of the Chemical Society* 101 (1912): 691.

48. Stanziani, "Information, Quality, and Legal Rules"; Stanziani, "Negotiating Innovation"; Stanziani, "La Mesure de la Qualité du Vin en France, 1871–1914"; Paul, *From Knowledge to Power*, chap. 5; Morris, *The Matter Factory: A History of the Chemistry Laboratory*, chaps. 9 and 10; Klein, *Experiments, Models, Paper Tools*; Simon, *Chemistry, Pharmacy, and Revolution in France, 1777–1809*; Tomic, *Aux origines de la chimie organique*; Tomic and Guillem-Llobat, "New Sites for Food Quality Surveillance in European Centres and Peripheries"; Tomic, "The Toxicological Laboratory of Paris during Jules Ogier's Direction 1883–1911."

49. The institute was one of four such state examination centers set up in Munich, Erlangen, Würzburg, and Speyer by royal decree in January 1884. For more information on German food research centers, see chap. 7.

50. Herz, "New Methods for Detecting Artificially Coloured Red Wines," cited in *Analyst* 1 (September 1886): 175.

51. K. Fleck, "A New Test for Picric Acid and Binitrocresol," *Repertorium der analytischen Chemie* 48 (1886), cited in *Analyst* 12 (January 1887): 16.

52. Curtman, "Test for Aniline Colours in Wines or Fruit Juices," *Zeitschrift für analytische Chemie* H4 (1887), cited in *Analyst* 12 (October 1887): 200–201.

53. Goldstein, *One Hundred Years of Medicine and Surgery in Missouri.*

54. Hofmann, *Liebigs Annalen der Chemie*, 144–214; Gautier, *Liebigs Annalen der Chemie*, 289. Isonitril, or isocyanide, is infamous for its obnoxious smell, a fact commented on by both Hofmann and Gautier. The Hofmann isonitril (now known as isocyanide) synthesis became, and still is, a key chemical test for primary amines based on their reaction with potassium hydroxide and chloroform. Hofmann had synthesized isonitrils (isomers of nitrites or isocyanides) through the action of chloroform and caustic potash on primary amines, while Gautier had isolated isonitrils (sometimes spelled "isonitriles" or, as Gautier called them, carbylamines) in a double decomposition reaction between silver cyanide and simple or compound ether.

55. Curtman, "Test for Aniline Colours in Wines or Fruit Juices," cited in *Analyst* 12 (October 1887): 200–201.

56. Klein, *Experiments, Models, Paper Tools*; Simon, *Chemistry, Pharmacy and Revolution in France, 1777–1809*; Tomic, *Aux origines de la chimie organique.*

57. *Analyst* 1 (1877): 186.

58. For more on spectroscopy, see Szabadváry, *History of Analytical Chemistry.*

59. *Analyst* 1 (1877): 186; Brears, *Jellies and Their Moulds*, 11.

60. *Analyst* 10 (October 1885): 179–81.

61. The toxicity of some of the new dyes and their ability to discriminately poison cells was recognized by Ehrlich, who developed chemotherapy as a means of treating diseases such as syphilis. Chemotherapy, the killing of cells using a range of toxic substances, is still a key treatment for cancer today. Schweitzer, "Ehrlich's Chemotherapy."

62. For more on the use of synthetic dyes in histology, see Travis, "Science as Receptor of Technology"; Cook, "Origins of Tinctorial Methods in Histology"; Collard and Collard, *The Development of Microbiology*; Wainwright, "The Use of Dyes in Modern Biomedicine"; Wainwright, "Dyes in the Development of Drugs and Pharmaceuticals"; Clark, Kasten, and Conn, *History of Staining.*

63. Schweissinger, "Microscopic Detection of Colouring Matters in Sausages," 53.

64. For more information and current regulation, see the web-sites of national and international food agencies, e.g., http://www.food.gov.uk/science/additives/foodcolours/#.U4LvXy_gLZs; http://www.efsa.europa.eu/en/topics/topic/foodcolours.htm (both accessed August 17, 2019).

65. French, *Antivivisection and Medical Science in Victorian Society*; Hamilton, *Animal Welfare and Anti-Vivisection 1870–1910*. For more on the work of Magendie and Bernal, see Porter, *The Greatest Benefit to Mankind*, chap. 11. See also Rupke, *Vivisection in Historical Perspective*; White, "Sympathy under the Knife."

66. Neswald, "Francis Gano Benedict's Reports."

67. Walter Fisher, evidence to the Parliamentary Select Committee, January 17, 1900, *Report of the Departmental Committee Appointed to Inquire into the Use of Preservatives and Colouring Matters in the Preservation and Colouring of Food*, 163–69.

68. Evidence of Professor John Attfield, February 9, 1900, ibid., 223–27. Clinicians such as the German physician Adolf Kaussman began to assess stomach contents using stomach tubes in the 1870s. For more information on stomach tubes and digestive experiments, see Rosenfeld, "Gastric Tubes, Meals, Acid, and Analysis"; Davenport, *A History of Gastric Secretion and Digestion*; Miller, *A Modern History of the Stomach*; Minard, "The History of Surgically Placed Feeding Tubes."

69. Conti and Bickel, "History of Drug Metabolism"; Chast, "Les Colorants, Outils Indispensables de la Révolution Biologique et Thérapeutique du XIXe Siècle."

70. Paul, *From Knowledge to Power*, 212; Cazeneuve, *Les Colorants de La Houille*; Schuchardt, "Ueber die Wirkungen des Anilins auf den thierischen

Organismus"; Filehne, "Ueber die Giftwirkungen des Nitrobenzols"; Turnbull, "On the Physiological and Medicinal Properties of Sulphate of Aniline, and Its Use in the Treatment of Cholera"; Engelhardt, *Beiträge zur Toxikologie des Anilin.*

71. Lesch, *Science and Medicine in France.*

72. Georges Bergeron and J. Cloüet, *Note sur l'innocuité Absolue des Mélanges Colorants à Base de Fuchsine Pure* (Rouen, 1876), cited in Weyl and Leffman, *The Coal-Tar Colors*, 23.

73. Paul, *From Knowledge to Power*, 213; Feltz and Ritter, *Etude Expérimentale de l'action de la Fuscine sur l'organisme*; Cazeneuve, *Les Colorants de La Houille.*

74. Cazeneuve and Lépine, "Les Couleurs de la Houille," cited in *Revue des sciences médicales en France et l'étranger* 31 (July 26, 1888); Cazeneuve and Lépine, "Sur les Effets Produits par l'ingestion et l'infusion Intravéneuse de Trois Colorants, Dérivés de La Houille." It is possible the food items singled out were those with the most persuasive producers, as Stanziani has argued in the case of French wine.

75. Weyl and Leffman, *The Coal-Tar Colors*, 24. Grandhomme was appointed to the Hoechst Colour Works in 1874 to assess the toxicity of raw materials used in the factories. Marquardt et al., *Toxicology*, 19.

76. International Labour Office, *Cancer of the Bladder among Workers in Aniline Factories.*

77. Hardy and Magnello, "Statistical Methods in Epidemiology"; Porter, *Karl Pearson*; Morabia, *A History of Epidemiologic Methods and Concepts.*

78. For more history on the work to determine the links between aniline and cancer, see Kennaway, "The Identification of a Carcinogenic Compound in Coal-Tar"; Lacassagne, "Kennaway and the Carcinogens"; Waller, "60 Years of Chemical Carcinogens"; Proctor, *The Nazi War on Cancer*; Case and Pearson, "Tumours of the Urinary Bladder in Workmen Engaged in the Manufacture and Use of Certain Dyestuff Intermediates in the British Chemical Industry." Meanwhile, more recent work suggests that incidences of bladder cancer in aniline factories may, in fact, have been attributable to naphthyamine, an aromatic amine used in the manufacture of azo dyes. See Kahl, "Aniline."

79. Weyl and Leffman, *The Coal-Tar Colors*, 25.

80. Ibid., 25.

81. Porter, *Trust in Numbers*; Desrosières, *The Politics of Large Numbers: A History of Statistical Reasoning*; Hardy and Magnello, "Statistical Methods in Epidemiology."

82. Porter, *Trust in Numbers*, preface.

83. Weyl and Leffman, *The Coal-Tar Colors*, 53.

84. Weyl's research included work on terpenes, the aromatic hydrocarbons produced in plants, and the development of what became known as Weyl's test, for color reactions in creatinine synthesized in the liver and kidneys. Weyl, "Über eine neue Reaction auf Kreatinin und Kreatin"; Delanghe and Speeckaert, "Creatinine Determination According to Jaffe."

85. Weyl and Leffman, *The Coal-Tar Colors*, 55. Dinitrocresol dyes were later marketed and used as insecticides and pesticides following work by the

German chemist Wilhelm von Miller, a professor of chemistry at Munich. See Vaupel, "Vom Teerfarbstoff zum Insektizid."

86. Hopwood, *Haeckel's Embryos*; Latour and Woolgar, *Laboratory Life*; Gradmann, "Experimental Life and Experimental Disease"; Todes, *Pavlov's Physiology Factory*.

87. Weyl and Leffman, *The Coal-Tar Colors*, 80.

88. Ibid., 85.

89. As mentioned earlier, Schultz produced an extensive description and classification of hundreds of the new synthetic dyes. Erdmann, Review of Schultz and Julius, *Tabellarische Übersicht der künstlichen organischen Farbstoffe.*

90. Weyl and Leffman, *The Coal-Tar Colors*, 94–96, 92.

91. Weyl claimed that Gnehm, using a preparation made by the Basel-based chemical manufacturer Binschedler & Busch, considered aurantia to be poisonous and argued that dyers using aurantia suffered swelling of the hands and arms, while Martius claimed that aurantia produced by his own company, AGFA, was not poisonous. The harmlessness of aurantia was also corroborated by experiments on rabbits by E. Salkowsky. Ibid., 94–96.

92. Ibid., 94–96.

93. Rowan, *Of Mice, Models, and Men*; Trevan, "The Error of Determination of Toxicity."

94. A.L. Winton, *Connecticut Agricultural Experiment Station Report* (Connecticut, 1901), 18.

95. Weyl and Leffman, *The Coal-Tar Colors*, 96.

96. Ibid., 115.

97. Fraenkel, *Die Arzneimittel Synthese auf Grundlage der Bezeitungen zwischen chemischen Aufbau und Wirkung.*

98. Schlacherl, *Fifth International Congress of Applied Chemistry.*

99. Lieber, *The Use of Coal Tar Colors in Food Products*, 150, cited in Hesse, *Coal-Tar Colors Used in Food Products*, 47; Hochheiser, "Synthetic Food Colors in the United States."

100. Chlopin, *Coal-Tar Dyes*, 114. An abstract of Chlopin's paper was printed in *Report of the Fifth International Congress of Applied Chemistry in Berlin* IV (1903): 169–72.

101. Chlopin, *Coal-Tar Dyes*, 219–21.

102. Ibid., 224.

103. Weber, "On the Behaviour of Coal-Tar Colors towards the Process of Digestion."

104. Gudeman, "Artificial Digestion Experiments."

105. Ibid., 1436.

106. Ibid.

107. For more about the secrecy in dye manufacturing in Germany at the time, see Pickering, "Decentering Sociology."

108. Hochheiser, "Synthetic Food Colors in the United States"; Hochheiser, "The Evolution of U.S. Food Colour Standards, 1913–1919"; Haynes, *American Chemical Industry*, vol. 2, 61.

109. Stilling, *Anilinfarbstoffe als Antiseptica*; Lehmann, *Methoden der praktischen Hygiene*.

110. *Forschungsberichte über Lebensmittel* 2 (1895): 181.

111. "Vorsätze der Schweizer Analytiker," 293.

112. *Food and Sanitation*, November 28, 1896, 574.

113. Bodewitz, Buurma, and de Vries, "Regulatory Science and the Social Management of Trust in Medicine," 251.

114. Velkar, *Markets and Measurements in Nineteenth-Century Britain*; Hunt, "The Ohm Is Where the Art Is"; Krislov, *How Nations Choose Product Standards and Standards Change Nations*; Buchwald, *Scientific Credibility and Technical Standards in 19th and Early 20th Century Germany and Britain*; Schaffer, "Metrology, Metrication, and Victorian Values."

115. Pinch, "Towards an Analysis of Scientific Observation."

116. For more on the history of standardization generally and electricity in particular see: Timmermans and Epstein, "A World of Standards but Not a Standard World"; Inkster, Gooday, and Sumner, *History of Technology*, vol. 28, special issue, *By Whose Standards? Standardization, Stability and Uniformity in the History of Information and Electrical Technologies*; Gooday, *The Morals of Measurement*; Schaffer, "Metrology, Metrication, and Victorian Values"; Schaffer, "Late Victorian Metrology"; Hughes, *Networks of Power*; Hirsh, *Technology and Transformation in the American Electric Utility Industry*; Finn, *The History of Electrical Technology*; Schmidt, *Coordinating Technology*.

117. Boddice, "Species of Compassion"; White, "The Experimental Animal in Victorian Britain."

118. Schaffer, "Astronomers Mark Time," 115.

119. Ibid., 118.

120. Collins, "Son of Seven Sexes," 34.

121. Collins, *Changing Order*; Gooding, Pinch, and Schaffer, *The Uses of Experiment*; Shapin and Schaffer, *Leviathan and the Air-Pump*; Pickering, *Science as Practice and Culture*.

122. For more on theories of analysis and synthesis as corresponding strategies in analytical chemistry, see Klein and Lefèvre, *Materials in Eighteenth-Century Science*, 115–66, 220.

123. Baeyer, *Ueber die chemische Synthese*; Partington, *A History of Chemistry*; Bensaude-Vincent, *A History of Chemistry*; Klein and Reinhardt, *Objects of Chemical Inquiry*.

124. Klein and Lefèvre, *Materials in Eighteenth-Century Science*; Klein and Reinhardt, *Objects of Chemical Inquiry*; Klein, *Experiments, Models, Paper Tools*; Bensaude-Vincent and Simon, *Chemistry*; Rocke, *Nationalizing Science*; Rocke, *Image and Reality*.

CHAPTER FIVE

1. Hamlin, *A Science of Impurity*; Waddington, "The Dangerous Sausage"; Waddington, *The Bovine Scourge*; Eyler, "The Epidemiology of Milk-borne Scarlet Fever"; Eyler, *Sir Arthur Newsholme and State Medicine, 1885–1935*;

Steere-Williams, "The Perfect Food and the Filth Disease"; Steere-Williams, "A Conflict of Analysis."

2. Hamlin, *A Science of Impurity*, 161. Hamlin notes that Hassall, like Edward Frankland, was not averse to "colouring facts" in order to "alarm the public mind."

3. British Government, *First Report from the Select Committee on the Adulteration of Food*, July 27. For more on the evidence given, see Steere-Williams, "The Perfect Food and the Filth Disease"; Royal Society of Chemistry, *The Fight Against Food Adulteration*.

4. Paulus, *Search for Pure Food*; Rioux, "Capitalist Food Production."

5. Paulus, *Search for Pure Food*; Rioux, "Capitalist Food Production."

6. British Government, The Report of the Committee of the Food Adulteration Act of 1872 (Read Committee) (London: H.M.S.O., 1874).

7. See Atkins, *Liquid Materialities*; French and Phillips, *Cheated Not Poisoned?*; Steere-Williams, "The Perfect Food and the Filth Disease." See also Hamlin, *A Science of Impurity*, for similar disputes over water quality; Rioux, "Capitalist Food Production."

8. Rioux, "Capitalist Food Production."

9. MacLeod, *Government and Expertise*; Hamlin, *Public Health*; Hamlin, *A Science of Impurity*; Eyler, *Sir Arthur Newsholme and State Medicine, 1885–1935*; Atkins, *Liquid Materialities*; Atkins, Lummel, and Oddy, *Food and the City*.

10. MacDonagh, *Early Victorian Government, 1830–1870*.

11. "Meeting of Public Analysts," 73.

12. Chirnside and Hamence, *Practising Chemists*; Hamlin, *A Science of Impurity*.

13. John Muter, *Analyst* 5 (February 1880): 15–16. Interestingly, just five years later, Muter noted that senior British government chemists and successful public analysts were paid more than their French counterparts. See chap. 7.

14. Burney, *Bodies of Evidence*; Gooday, "Liars, Experts, and Authorities."; Hamlin, *A Science of Impurity*.

15. Cameron, "President's Address."

16. Hehner, "President's Address," 8; "Remuneration of Public Analysts," 34.

17. For a full account of this dispute and the acrimonious falling out between Hassall and Letheby, see Steere-Williams, "The Perfect Food and the Filth Disease."

18. Atkins, *Liquid Materialities*, 180.

19. "Food and Drugs (Adulteration) Bill," 778.

20. Chirnside and Hamence, *Practising Chemists*, 3.

21. For more on changing sites and concepts of science in the nineteenth century, see Fyfe and Lightman, *Science in the Marketplace*; Cahan, *From Natural Philosophy to the Sciences*. Also on the discipline and status of chemists, see Inkster and Morrell, *Metropolis and Province*; Bud and Roberts, *Science Versus Practice*; Berman, *Social Change and Scientific Organization*.

22. Homburg, "The Rise of Analytical Chemistry."

23. Brock, *William Crookes (1832–1919) and the Commercialization of*

Science. The industrial chemist Ivan Levenstein similarly founded *The Chemical Review* (1871–1891), a monthly journal for "manufacturing chemists and druggists, dyers, printers, bleachers, sizers, paper-makers and strainers etc," just eight years after his arrival in England from Germany. Another chemist with experience at the more popular end of the Victorian media was leading public analyst Bernard Dyer, one of the founders of the Society of Public Analysts, whose father was news editor of the *Daily News*. *Analyst* 890 (1950): 240.

24. Burns, *Bad Whisky*.

25. The Glasgow MP's interest in food adulteration often resulted in his being mistaken for Dublin's public analyst Sir Charles Alexander Cameron, who was appointed president of the SPA in 1994 and whose name he shared.

26. "The Public and 'Public Analysts,'" 155–56. For more on the low status of public analysts and their insecurity, see French and Phillips, *Cheated Not Poisoned?*, 39.

27. For more on the role of consumers during this period, see Daunton and Hilton, *The Politics of Consumption*; French and Phillips, "Sophisticates or Dupes?"; Kassim, "The Co-operative Movement and Food Adulteration in the Nineteenth Century"; Trentmann, *Empire of Things*.

28. Volume 1 of the *Analyst* (1876) refers to several articles in the *Local Government Chronicle* and *The Medical Examiner* claiming the Society of Public Analysts was in disarray with the resignation of various officials.

29. For more information on the multitude and quantities of additives supplied by chemists to the whisky and beer trades, see Burns, *Bad Whisky*.

30. The consulting chemists of the nineteenth century are not dissimilar to Ursula Klein's seventeenth-century "hybrid experts," chemists who worked for both the state and industry as well as holding academic posts. Klein and Spary, eds., *Materials and Expertise in Early Modern Europe*.

31. For examples of vitriolic disputes between chemists, see Hamlin, *A Science of Impurity*.

32. "Organization Amongst Chemists," *Analyst* 2, 19 (1877): 109–11; "Editorial," *Analyst* 2, 22 (1878): 171–72; "Notes of the Month," *Analyst* 2, 23 (1878): 205–6; "Notes of the Month," *Analyst* 3, 29 (1878): 316.

33. "Pure Food."

34. "Report of the Annual Meeting."

35. Hamlin, *A Science of Impurity*.

36. British Government, *The Report of the Committee of the Food Adulteration Act of 1872 (Read Committee)*, QQ 5589, 5861, 5600.

37. Correspondence between Richard Bannister of Somerset House and Alfred Allen in the *Analyst* 19 (1884): 231–40, cited in Atkins, *Liquid Materialities*, 96.

38. Hammond and Egan, *Weighed in the Balance*. For an early history of customs and excise in Britain, see Ashworth, *Customs and Excise*.

39. Hammond and Egan, *Weighed in the Balance*.

40. Ibid., 13–30, 35.

41. Excise Office, *Report of the Principal Chemist*, 36.

42. Hammond and Egan, *Weighed in the Balance*.

43. Ibid.

44. British Government, *The Report of the Committee of the Food Adulteration Act of 1872 (Read Committee)*, 95.

45. Russell, *Edward Frankland*, 376.

46. Hammond and Egan, *Weighed in the Balance*, 88.

47. Fox, *Dye-Makers of Great Britain 1856–1976*, 4. Frankland, along with Hofmann, William Crookes, Henry Letheby, and several other notable chemists appeared for the plaintiff. Wanklyn, meanwhile, appeared for the defendant Levenstein, claiming that Levenstein had originally made the dye in London using, under license, a process Wanklyn had developed in Heidelberg. The case was dissolved on the grounds that no infringement of the patent had been proved.

48. For more on Wanklyn and his controversial dispute with Frankland, see Hamlin, *A Science of Impurity*.

49. "Editorial," *Food, Drugs and Drink*, 2, 27 (1893): 1. For examples of more comments in the press, see Steere-Williams, "Lacteal Crises—Debates over Milk Purity in Victorian England."

50. Burns, *Bad Whisky*, 87.

51. Steere-Williams, "A Conflict of Analysis."

52. Hamlin, "Scientific Method and Expert Witnessing."

53. "The Whole Duty of a Chemist."

54. "Editorial," *The Times*, April 4, 1882, 9, col. C; Huxley, *The Life and Letters of Thomas Henry Huxley*, 255, 257–58; *The Spectator* 67 (1891): 525, cited in Gooday, "Liars, Experts, and Authorities," 435.

55. Golan, *Laws of Men and Laws of Nature*; Gooday, "Liars, Experts, and Authorities."

56. Gooday, "Liars, Experts, and Authorities"; *Nature* 33 (Nov 16, 1885): 73–77; *Nature* 33 (Dec 3, 1885): 99; *The Spectator* 67 (1891): 525; *Chemical News*, 5 (1862): 183; *Chemical News* 29 (1874): 216, 249: *Chemical News* 47 (1886); 107; *Chemical News* 53 (1886): 1–2, 39, 72. For more on the history of chemists and other scientists acting as expert witnesses in court, see Hamlin, "Scientific Method and Expert Witnessing"; Kargon, "Expert Testimony in Historical Perspective"; Landsman, "Of Witches, Madmen, and Products Liability."

57. For more about different forums for assessing risk and quality, see Callon, Lascoumes, and Barthe, *Acting in an Uncertain World*.

58. Atkins, *Liquid Materialities*, chap. 4; Hildebrandt, "The Trial of the Expert."

59. "The Somerset House Court of Appeal"; "Report of Meeting."

60. "Editorial and Exchange of Letters."

61. Atkins, *Liquid Materialities*, chap. 7; Porter, *Trust in Numbers*; Gooday, *The Morals of Measurement*; Velkar, *Markets and Measurements in Nineteenth-Century Britain*.

62. Stanziani, *Rules of Exchange*; Stanziani, "Negotiating Innovation."

63. Hamlin, *A Science of Impurity*, 224.

64. For more on the government chemists' experiments on different food types, see Atkins, *Liquid Materialities*; Chirnside and Hamence, *Practising Chemists*; Hammond and Egan, *Weighed in the Balance*.

65. Hammond and Egan, *Weighed in the Balance*. For more on disputes between pubic analysts and Somerset House, see Chirnside and Hamence, *Practising Chemists*; Atkins, *Liquid Materialities*; Steere-Williams, "The Perfect Food and the Filth Disease"; Atkins, "Sophistication Detected." See also Steere-Williams, "A Conflict of Analysis."

66. Atkins, *Liquid Materialities*, 70; Hehner, "On the Relation between the Specific Gravity, Fat and Solids-Non-Fat in Milk, upon the Basis of the Society of Public Analysts' Method."

67. Hehner, "Abstract of the Work of the Milk Committee," 3.

68. Ibid.

69. Bell, "Food Adulteration and Analysis."

70. Atkins, *Liquid Materialities*, chap. 2; Pickering, *The Mangle of Practice*.

71. Allen, "Response."

72. Hehner, "Abstract of the Work of the Milk Committee," 3.

73. Burnett, *Plenty and Want*.

74. James Bell, "Letter to Robert McAlley, Public Analyst for Falkirk" (Letter, London, March 1877), DSIR 26/118, National Archives, cited by Steere-Williams, "A Conflict of Analysis," 291.

75. See Dyer et al., *The Society of Public Analysts and Other Analytical Chemists*; Chirnside and Hamence, *Practising Chemists*; Atkins, *Liquid Materialities*.

76. Steere-Williams, "A Conflict of Analysis," 296.

77. Hammond and Egan, *Weighed in the Balance*.

78. "Proceedings of the Society of the Public Analysts," 110.

79. Dyer et al., *The Society of Public Analysts and Other Analytical Chemists*.

80. O'Rourke, "British Trade Policy in the 19th Century."

81. Spencer, "From Freedom to Bondage," introduction.

82. Bannister, "The Food of the People."

CHAPTER SIX

1. Cameron, "Address given at the Meeting for the Society of Public Analysts," 175.

2. *Transactions of the Royal Academy of Medicine in Ireland* 9, 1 (December 1891): 447–49; *Lancet*, 12 March, 1891, 667.

3. Ibid.

4. Howe and Howe, *Free Trade and Liberal England, 1846–1946*.

5. For more on consumerism and the political economy in this period, see Trentmann and Daunton, *Worlds of Political Economy*; Hilton, *Consumerism in Twentieth-Century Britain*; Okun, "Fair Play in the Marketplace"; Baudrillard, *The Consumer Society*; Benson, *The Rise of Consumer Society in Britain, 1880–1980*; Chatriot, Chessel, and Hilton, *The Expert Consumer*; Howe and Howe, *Free Trade and Liberal England, 1846–1946*.

6. Oddy, "Food Quality in London and the Rise of the Public Analyst," 95–96; Collins and Oddy, "The Centenary of the British Food Journal."

7. Natrajan, Dyer, and Clayton, "Obituary Notices."

8. Margarine, invented at the request of a government, has arguably also

been one of the most regulated and controlled food commodities. Stuyvenberg, ed., *Margarine*.

9. British Government, *Report of the Departmental Committee Appointed to Inquire into the Use of Preservatives and Colouring Matters in the Preservation and Colouring of Food*, paragraph 57. The historian Heiko Stoff has done some interesting work on butter-yellow controversies in the twentieth century. See Stoff, *Gift in der Nahrung*.

10. Scott Elder, *Appeal Cases Under the Sale of Food & Drugs Acts, 1875 & 1879, and the Margarine Act, 1887*; Bartley, *Adulteration of Food: Statutes and Regulations, Including the Food and Drugs (Adulteration) Act, 1928, and Dealing with Coffee, Tea, Bread, Butter, Milk, Margarine, Margarine Cheese, Milk-Blended Butter and All Other Foods, and Drugs*.

11. "A Bill to Regulate the Importation, Manufacture and Sale of Butter Substitutes," 100.

12. Evidence of Robert Gibson, speaking as the representative of the South of Ireland Butter Merchants' Association, February 7, 1900, British Government, *Report of the Departmental Committee Appointed to Inquire into the Use of Preservatives and Colouring Matters in the Preservation and Colouring of Food*, 215, 218.

13. "The Grocers' Federation: Report on Food Legislation," 45.

14. "Public Analysts and the State," 9.

15. Ibid.

16. Ibid.

17. August Dupré, evidence to the Select Committee, January 1900, British Government, *Report of the Departmental Committee Appointed to Inquire into the Use of Preservatives and Colouring Matters in the Preservation and Colouring of Food*, 202.

18. Otto Hehner, evidence to the Select Committee, January 19, 1900, ibid., 193.

19. Trentmann and Taylor, "From Users to Consumers," 53–59.

20. See chap. 4 for more on this debate.

21. Levenstein, *Revolution at the Table*, 32–33. See Spary, *Feeding France*, 285–314, for a nuanced description of how beet sugar won political and chemical support in France as a substitute for cane sugar, while the chemical similarity between the two types of sugar made it difficult for chemists to distinguish between them. Meanwhile, the debate over the two types of sugars continues today. Greer, "Britain Doesn't Need Beet Sugar"; Morgan, "Sugar, Sugar."

22. Cassal, "On Dyed Sugar."

23. Ibid., 145.

24. Ibid., 146–47.

25. Ibid., 147.

26. Ibid.

27. "Birmingham Court Case."

28. Ibid., 143.

29. *The Times*, September 20, 1923, cited in French and Phillips, *Cheated Not Poisoned?*, 30.

30. For more about the use of chloride of tin, see the sugar debate reported in Cassal, "On Dyed Sugar."

31. Warnford Lock, Newlands, and Newlands, *Sugar*.

32. Hugill, *Sugar and All That*; Bender, *A Dictionary of Food and Nutrition*, 515.

33. Golan, *Laws of Men and Laws of Nature*; Hamlin, "Scientific Method and Expert Witnessing."

34. Jones, *Expert Witnesses*.

35. Paulus, *Search for Pure Food*, 43; Rioux, "Capitalist Food Production."

36. British Government, *Report of the Departmental Committee Appointed to Inquire into the Use of Preservatives and Colouring Matters in the Preservation and Colouring of Food*.

37. MacLeod, *Government and Expertise*; Jones, *Expert Witnesses*.

38. Otto Hehner, evidence to the Select Committee, January 19, 1900, British Government, *Report of the Departmental Committee Appointed to Inquire into the Use of Preservatives and Colouring Matters in the Preservation and Colouring of Food*, 191–94.

39. Cassal, "On Dyed Sugar," 147.

40. Otto Hehner, evidence to the Select Committee, January 19, 1900, British Government, *Report of the Departmental Committee Appointed to Inquire into the Use of Preservatives and Colouring Matters in the Preservation and Colouring of Food*, 191–94.

41. British Government, *Report of the Departmental Committee Appointed to Inquire into the Use of Preservatives and Colouring Matters in the Preservation and Colouring of Food*, paragraphs 63, 65.

42. "Bannister's Evidence," 283.

43. British Government, *Report of the Departmental Committee Appointed to Inquire into the Use of Preservatives and Colouring Matters in the Preservation and Colouring of Food*.

44. Walter Fisher, evidence to the Parliamentary Select Committee, January 17, 1900, ibid., 163–69.

45. Alexander Wynter Blyth, evidence to the Parliamentary Select Committee, January 17, 1900, ibid., 115–20.

46. Charles Cassal, evidence to Parliamentary Select Committee, January 15, 1900, ibid., 129–36.

47. Ibid.

48. Ibid.

49. Ibid., paragraphs 126–35.

50. Ibid., paragraphs 126, 128, 130, 131.

CHAPTER SEVEN

1. Atkins and Stanziani, "From Laboratory Expertise to Litigation"; Atkins, Lummel, and Oddy, *Food and the City*; Dessaux, "Chemical Expertise and Food Market Regulation in Belle-Epoque France."

2. Stanziani, "Municipal Laboratories and the Analysis of Foodstuffs in France under the Third Republic"; Tomic and Guillem-Llobat, "New Sites for Food Quality Surveillance in European Centres and Peripheries."

3. Stanziani, "Information, Quality, and Legal Rules," 273.

4. Gautier, "The Fraudulent Colouration of Wines," 110.

5. Dessaux, "Chemical Expertise and Food Market Regulation in Belle-Epoque France."

6. Zylberman, "Making Food Safety an Issue."

7. Atkins and Stanziani, "From Laboratory Expertise to Litigation."

8. Huet-Desaunay, *Le Laboratoire Municipal et les Falsifications ou Recueil des Lois et Circulaires Concernant la Vente des Produits Alimentaires et Hygiene Publique* (Paris, 1890); Paul, *From Knowledge to Power*, 211–20; Dessaux, "Chemical Expertise and Food Market Regulation in Belle-Epoque France."

9. Muter, "On the Processes and Standards of Food Analysis in Use at the Municipal Laboratory of the City of Paris," 144.

10. Ibid., 145.

11. Denys Cochin, *Revue des Deux Mondes*, June 15, 1883, cited in "Adulteration in Paris," 239–40.

12. Charles, *Documents sur les Falsifications des Matières Alimentaires et sur les Travaux du Laboratoire Municipal: Rapport à Monsieur le Préfet de Police: Deuxième Rapport* (Paris, 1885), cited in Paul, *From Knowledge to Power*, 215.

13. Dessaux, "Chemical Expertise and Food Market Regulation in Belle-Epoque France"; Coppin and High, *The Politics of Purity*.

14. Hierholzer, "Searching for the Best Standard"; Dessaux, "Chemical Expertise and Food Market Regulation in Belle-Epoque France"; Spiekermann, "Redefining Food."

15. Testimony of Georges Berry in *Journal Officiel*, Assemblée Nationale (1904), 2356, 2358, cited in Dessaux, "Chemical Expertise and Food Market Regulation in Belle-Epoque France," 354.

16. Bergeron, *Rapport sur les proprieties toxiques de la fuchsine non arsenicale* (Recueil des travaux du Comité consultative d'hygiène, 7, 321), cited in Cazeneuve, *Les Colorants de la Houille*, 14; Henri Fauvel obituary, *Lancet*, 1886, 1001. For more on Proust, see Bogousslavsky, *Following Charcot*, 68. In 1885, Proust, the father of novelist Marcel Proust, founded the International Office of Hygiene, the predecessor to the World Health Organisation. For works by Bussy, see Wehefritz, *Bibliographie zur Geschichte der Chemie und chemischen Technologie. 17. bis 19. Jahrhundert*, 664.

17. Cazeneuve and Lépine, "Les Couleurs de la houille," cited in *Revue des sciences médicales en France et l'étranger*, vol. 31, July 26, 1888, 517; Cazeneuve and Lépine, "Sur les effets produits par l'ingestion et l'infusion intraveneuse de trois colorants, derives de la houille"; *Bulletin de l'Académie de Médecine* 16, 2nd s. (1886): 310. It is possible the food items singled out were those with the most persuasive producers, as Stanziani has argued in the case of French wine.

18. For more on legislation in France and other countries, see table 3.1.

19. Cazeneuve, *Les Colorants de la Houille*.

20. Lieber, *The Use of Coal Tar Colors in Food Products*.

21. Atkins, "The Material Histories of Food Quality and Composition"; Gade, "Tradition, Territory, and Terroir in French Viniculture."

22. Atkins and Stanziani, "From Laboratory Expertise to Litigation."

23. Teuteberg, "Adulteration of Food"; Hierholzer, "Searching for the Best Standard"; Ellerbrock, "Lebensmittelqualität vor dem Ersten Weltkrieg"; Hierholzer, *Nahrung nach Norm*; Grüne, *Anfänge staatlicher Lebensmittelueberwachung in Deutschland*.

24. *Denkschrift über die Aufgaben und Ziele, die das kaiserliche Gesundheits-Amt sich gestellt hat*, Berlin, 1878, 2, cited in Hierholzer, "Searching for the Best Standard," 305.

25. Buchka, *Die Nahrungsmittelgesetzgebung im Deutschen Reiche*.

26. König and Juckenack, "Preußen."

27. Ellerbrock, "Lebensmittelqualität vor dem Ersten Weltkrieg."

28. "Verschiedene Arbeiten zur Untersuchung der künstlichen Färbung von Würsten in Deutschland" (Berlin), Imperial Health Office R86/2255, Deutsche Archive.

29. Ellerbrock, "Lebensmittelqualität vor dem Ersten Weltkrieg"; Hierholzer, *Nahrung nach Norm*; Grüne, *Anfänge staatlicher Lebensmittelüberwachung in Deutschland*.

30. Bujard and Baier, *Hilfsbuch für Nahrungsmittelchemiker*; Hasterlik, *Die praktische Lebensmittelkontrolle*; Neufeld, *Der Nahrungsmittelchemiker als Sachverständiger*; Rupp, *Die Untersuchung von Nahrungsmitteln*.

31. *Vereinbarungen zur einheitlichen Untersuchung und Beurtheilung von Nahrungs- und Genußmitteln sowie Gebrauchständen für das Deutsche Reich*; Hierholzer, "Searching for the Best Standard."

32. Specialist food industry journals included *Zeitschrift fur Untersuchung der Nahrungs- und Genussmittell wowie Gebrauchsgegenstände* (ZUNG), *Deutsche Nahrungsmittel-Rundschau, Konserven-Zeitung, Monatsschrift für die volkswirtschaftlichen, gesetzgeberischen und kommerziellen Interessen der Margarine-Industrie, Pflanzenfett- und Seisolbereitung*.

33. Ellerbrock, "Lebensmittelqualität vor dem Ersten Weltkrieg," 141.

34. Neufeld, *Der Nahrungsmittelchemiker als Sachverständiger*, 4.

35. *Bericht über den XIV Internationale Kongress*, 321; Ellerbrock, "Lebensmittelqualität vor dem Ersten Weltkrieg," 141.

36. Neufeld, *Der Nahrungsmittelchemiker als Sachverständiger*; König, *Chemie der menschlichen Nahrungs- und Genussmittel*, vol. 2, 462.

37. Ellerbrock, "Lebensmittelqualität vor dem Ersten Weltkrieg," 141.

38. Ibid.

39. Kerp, *Nahrungsmittelchemische Tagesfragen*, 191; *Deutsches Nahrungsmittelbuch*, 12; Todes, *Pavlov's Physiology Factory*.

40. Weyl and Leffman, *The Coal-Tar Colors*, xi.

41. "Various Papers Relating to the Use of Eosin in Food" (Berlin), Imperial Health Office papers, BA R86/2297, Deutsche Archive, Berlin.

42. For more on eosin and its discovery, see Reinhardt, *Heinrich Caro and the Creation of Modern Chemical Industry*.

43. Murmann, "Knowledge and Competitive Advantage"; Beer, *The Emergence of the German Dye Industry*; Beer, "Coal Tar Dye Manufacture and the Origins of the Modern Industrial Research Laboratory."

44. *Berliner Lokal-Anzeiger Morgenblatt*, December 15, 1909.

45. Letter to der Staatsekretär des Innern from the president of the KGA,

May 14, 1910, Imperial Health Office papers, BA R86/2297, Deutsche Archive, Berlin.

46. Rost, *Pharmakologische Untersuchung des Eosins*, Imperial Health Office papers, BA R86/2297, Deutsche Archive, Berlin.

47. Letter and article sent to RGA from Dr. Fritz Schanz, October 9, 1915; Fritz Schanz, "Sonnenstich-Hitzschlag," *München Medizinische Wochenschrift*, 1915, 29; Fritz Schanz, "Die Wirkungen des Lichtes auf die lebenden Organismen," *Biochemische Zeitschrift*, 71. All cited items are in Imperial Health Office papers, BA R86/2297, Deutsche Archive, Berlin.

48. E. Rost, "On the Physiological Affect and Toxicology of Eosins," a special supplement in *Medizinische Klinik*, 1915, 36, Imperial Health Office papers, BA R86/2297, Deutsche Archive, Berlin.

49. Letter to the Direktorium der Reichsgetreidestelle from Geheimer Rat des Reichs-Gesundheitrates President des Königlichen Landes-Gesundheitsamtes, November 9, 1915, Imperial Health Office papers, BA R86/2297, Deutsche Archive, Berlin.

50. *Medizinische Klinik: Wochenschrift für praktische Ärtze* 51 (1915) (offprint); letters from President Dr. Bumm to die Reichsgetreidestelle Geschäftsableitung, March 1, 1916. Both cited items are in Imperial Health Office papers, BA R86/2297, Deutsche Archive, Berlin.

CHAPTER EIGHT

1. Wiley in his own autobiography together with his official biographer, Oscar Anderson, were instrumental in creating this early image of Wiley. Wiley, *The History of a Crime*; Anderson, *The Health of a Nation*.

2. Levenstein, *Revolution at the Table*; Levenstein, *Fear of Food*; Coppin and High, *The Politics of Purity*, 3, chap. 3; Anderson, *The Health of a Nation*; Wood, "The Strategic Use of Public Policy"; Young, *Pure Food*; Cohen, *Pure Adulteration*.

3. Harvey W. Wiley's testimony on February 26, 1906, in US Government, *Hearings . . . on the pure-food bills*, 317.

4. Walter H. Williams's testimony on February 14, 1906, in *Hearings . . . on the pure-food bills*, 23.

5. Harvey W. Wiley's testimony on February 26, 1906, in *Hearings . . . on the pure-food bills*, 240.

6. *Oleomargarine and Butterine*, 3.

7. For more on this, see Coveney, *Food, Morals, and Meaning*. The connection between food, morals, and urban decay frequently featured in Victorian novels according to Hyman, *Making a Man*.

8. Miller, "Public Choice at the Dawn of the Special Interest State," 115; Ball and Lilly, "The Menace of Margarine."

9. "Progress and Butter."

10. Miller, "Public Choice at the Dawn of the Special Interest State", 82. For an in-depth study of margarine legislation in the US pre-1930 and the economic, cultural, and chemical comparisons with butter, see Snodgrass, *Margarine as a Butter Substitute*.

11. Ball and Lilly, "The Menace of Margarine," 492; Douglas, *Purity and Danger*.

12. For an example of how "ritually significant foods" are created, see Boisard, *Camembert*.

13. Ibid.; Burns, "Bogus Butter," abstract.

14. Martin, "Detection of Artificial Colouring Matter in Butter, Oleomargarine, Fats, Oils Etc"; Leeds, "Methods for Separating Out Colours in Butter, Imitation Butter."

15. For more on the history of margarine in Canada, see Heick, *A Propensity to Protect*.

16. *Collins v. New Hampshire*, 171 US 30 (1898), cited in Burrows, "Palette of Our Palates," 397.

17. Stuyvenberg, *Margarine*, 281–319.

18. Reynolds, "Defining Professional Boundaries"; Coppin and High, *The Politics of Purity*, 39.

19. Marcus, "Setting the Standard"; Coppin and High, *The Politics of Purity*, chap. 3.

20. Harvey W. Wiley's testimony on February 16, 1906, in *Hearings . . . on the pure food bills*, 237.

21. Spiekermann, "Redefining Food."

22. Haynes, *American Chemical Industry*, vol. 2, 61; Bernhard Hesse, *Oil, Paint, and Drug Reporter*, February 9, 1917, 9, cited in Hochheiser, "Synthetic Food Colors in the United States."

23. Hesse, *Coal-Tar Colors Used in Food Products*; Hochheiser, "Synthetic Food Colors in the United States."

24. Hesse, *Coal-Tar Colors Used in Food Products*, 15.

25. Ibid.; Hesse, "The Industry of the Coal-Tar Dyes."

26. Hesse, *Coal-Tar Colors Used in Food Products*, 16–18.

27. Hochheiser, "Synthetic Food Colors in the United States."

28. USDA, Office of the Secretary, FID 76, "Dyes, Chemicals and Preservatives in Food," July 13, 1907.

29. Hochheiser, "Synthetic Food Colors in the United States," 52; *Handbook of U.S. Colorants*, 10–12; USDA, Office of the Secretary, FID 117, "The Use of Certified Colors," May 3, 1910.

30. Miller, "Food Colours," 347.

31. Wiley, *The History of a Crime*.

32. Harvey W. Wiley's testimony, February 26, 1906 in *Hearings . . . on the pure-food bills*, 1906, 244.

33. Ibid., 362.

34. Wiley, *The History of a Crime*; Wiley and Pierce, *1001 Tests of Foods, Beverages and Toilet Accessories, Goods and Otherwise*.

CONCLUSION

1. Blythman, *Swallow This*, 3.

2. Schug et al., "Endocrine Disrupting Chemicals and Disease Susceptibility"; Hanan et al., "Toxic Effects of Some Synthetic Food Colorants and/

or Flavor Additives on Male Rats"; Arnold et al., "Artificial Food Colors and Attention-Deficit/Hyperactivity Symptoms"; Bernhardt, Rosi, and Gessner, "Synthetic Chemicals as Agents of Global Change"; Boudia et al., "Residues."

3. Collins, "Son of Seven Sexes"; Collins, *Changing Order*; Gooding, Pinch, and Schaffer, *The Uses of Experiment*; Shapin and Schaffer, *Leviathan and the Air-Pump*; Pickering, *Science as Practice and Culture*.

4. For more on the application of different chemical theories and practices, see Klein and Lefèvre, *Materials in Eighteenth-Century Science*; Klein and Reinhardt, *Objects of Chemical Inquiry*; Klein, *Experiments, Models, Paper Tools*; Bensaude-Vincent and Simon, *Chemistry*; Rocke, *Nationalizing Science*; Rocke, *Image and Reality*.

5. Petrick, "The Industrialization of Food"; White, "Chemistry and Controversy"; White Junod, "The Chemogastric Revolution"; Gratzer, *Terrors of the Table*.

6. French and Phillips, *Cheated Not Poisoned?*; Stanziani, "Information, Quality, and Legal Rules."

7. Cantor et al., *Science in the Nineteenth-Century Periodical*.

8. Otterloo, "The Development of Public Distrust of Modern Food Technology in the Netherlands."

9. Pinch, "Towards an Analysis of Scientific Observation."

10. Chang, "Circularity and Reliability in Measurement"; Collins, *Changing Order*; Gooday, *The Morals of Measurement*; Gooding, Pinch, and Schaffer, *The Uses of Experiment*; Schaffer, "Metrology Metrication, and Victorian Values"; Shapin and Schaffer, *Leviathan and the Air-Pump*.

11. Weyl, *Handbuch der Hygiene*.

12. Ibid.

13. Travis, *The Rainbow Makers*; Homburg, Travis, and Schröter, *The Chemical Industry in Europe*; Pickstone, *Ways of Knowing*; Gee, "Amusement Chests and Portable Laboratories."

14. Johnston, *A History of Light and Colour Measurement*; Cochrane, *Measures for Progress*.

15. Spiekermann, "Redefining Food"; Dessaux, "Chemical Expertise and Food Market Regulation in Belle-Epoque France"; Guillem-Llobat, "The Sugar Industry."

16. Hamlin, *A Science of Impurity*; Hamlin, *Public Health*.

17. Blythman, *Swallow This*; Patel, *Stuffed and Starved*; Pollan, *Cooked*.

18. Beck, *Risk Society*; Collins and Evans, *Rethinking Expertise*; Giddens, *The Consequences of Modernity*.

19. Collins and Evans, "The Third Wave of Science Studies"; Jasanoff, "Breaking the Waves in Science Studies"; Ash, "Expertise and the Early Modern State."

20. Weyl, *Handbuch der Hygiene*; Weyl, *The Coal-Tar Colours*.

21. Wiley, *The History of a Crime*.

22. Lieber, *The Use of Coal Tar Colors*.

23. Gooday, "Liars, Experts, and Authorities."

24. Rioux, "Capitalist Food Production."

25. Blythman, *Swallow This*, 157.

26. Wiley, *The History of a Crime.*

27. Homburg, Travis, and Schröter, *The Chemical Industry in Europe*, introduction.

Bibliography

ARCHIVE MATERIAL

Deutsche Archive, Advertisements, Imperial Health Office, R86/2255, Berlin, Germany.

Deutsche Archive, Various Papers Relating to an Investigation into the Artificial Colouring of Sausages in Germany, Imperial Health Office R86/2255, Berlin.

Deutsche Archive, Various Papers Relating to the Use of Eosin in Food, Imperial Health Office R86/2297, Berlin.

Huntley and Palmer Archive, Correspondence, HP/143, Museum of English Rural Life, Reading.

National Archives, Correspondence, Government Laboratory, DSIR 26/118, London.

Rowntree Archive, Chemical Department, R/DT/CC/6, Borthwick Institute, York.

Rowntree Archive, Correspondence, H. I. Rowntree & Co. The Cocoa Works 1887–1904. HIR/1/15. Rowntree Archives, Borthwick Institute, York.

Rowntree Archive, J. W. Rowntree, Private Experimental Notebooks, H. I. Rowntree & Co. The Cocoa Works 1887–1904. HIR/7b/10, Borthwick Institute, York.

Rowntree Archive, Rowntree Recipe Books, H. I. Rowntree & Co. The Cocoa Works 1887–1904. HIR/7B/2, Borthwick Institute, York.

US National Archives, General Correspondence, Bureau of Chemistry, Record Group 97, Washington DC, US.

Williams Bros Archives, Various Papers, Hounslow Library, London.

GOVERNMENT REPORTS

British Government. *First Report from the Select Committee on the Adulteration of Food.* London: H.M.S.O., 1855.

British Government. *The Report of the Committee of the Food Adulteration Act of 1872 (Read Committee).* London: H.M.S.O., 1874.

British Government. *Report of the Departmental Committee Appointed to Inquire into the Use of Preservatives and Colouring Matters in the Preservation and Colouring of Food: Together with Minutes of Evidence, Appendices and Index.* London: H.M.S.O., 1901.

US Government, *Hearings before the Committee on Interstate and Foreign Commerce of the House of Representatives (February 13–27, 1906) on the pure-food bills H.R. 3044, 1527, 7018, 12071, 13086, 13853, and 13859, for preventing the adulteration, misbranding, and imitation of foods, beverages, candies, drugs, and condiments in the District of Columbia and the territories, and for regulating interstate traffic therein, and for other purposes* (Washington: Government Printing Office, 1906).

PRIMARY PRINTED SOURCES

"Accidents and Offences." *John Bull & Britannia*, 2844 (1875): 402.

Accum, Friedrich Christian. *A Treatise on Adulterations of Food and Culinary Poisons: Exhibiting the Fraudulent Sophistications of Bread, Beer, Wine, Spirituous Liquors, Tea, Coffee, Cream, Confectionery, Vinegar, Mustard, Pepper, Cheese, Olive Oil, Pickles and Other Articles Employed in Domestic Economy and Methods of Detecting Them.* London: Printed by J. Mallett and sold by Longman, Hurst, Rees, Orme, and Brown, 1820.

"Adulteration in Paris." *Analyst* 8 (November 1883): 239–40.

"Advertisement." *Leeds Mercury*, September 27, 1865.

"Advertisement for Eno's." *Pall-Mall Gazette* 7598 (1889).

Allen, Alfred H. "Response." *Analyst* 19 (March 1894): 56.

"Aniline Colours." *Bradford Observer*, 1892 (January 14, 1869): 3.

Arata, Pedro. "Detection of Colouring Matters in Wine." *Anales de La Sociedad Científica Argentina* 19 (1885): 140.

Baeyer, Johann Friedrich Wilhelm Adolf von. *Ueber die chemische Synthese: Festrede, etc.*München: K B Akademie, 1878.

Bannister, Richard. "The Food of the People." In J. Samuelson, *The Civilisation of Our Day: Essays*, chapter 2. London: S. Low, Marston and Co., 1896.

"Bannister's Evidence." *Analyst* 19 (December 1894): 283.

Bartley, Douglas C. *Adulteration of Food: Statutes and Regulations, Including the Food and Drugs (Adulteration) Act, 1928, and Dealing with Coffee, Tea, Bread, Butter, Milk, Margarine, Margarine Cheese, Milk-Blended Butter and All Other Foods, and Drugs.* London: Stevens, 1929.

Baum, L. Frank. *American Fairytales.* Chicago: George M. Hill & Co., 1901.

Baum, L. Frank. *The Art of Decorating Dry Goods Windows and Interiors: A Complete Manual of Window Trimming, Designed as an Educator in all the details of the Art, According to the Best Accepted Methods, and Treating Fully Every Important Subject* Chicago: Snow Window Publishing Co., 1900.

Baum, L. Frank. *The Wonderful Wizard of Oz.* Chicago: George M. Hill & Co., 1900.

"Beautiful Tar: Song of an Enthusiastic." *Punch*, September 15, 1888, 123.

Belar, A. "Detection of Foreign Colouring Matters in Red Wines." *Zeitschrift für analytische Chemie* 35 (1896): 322–23.

Bell, James. "Food Adulteration and Analysis." *Analyst* 9 (1884): 133–47.

Bergeron, Georges, and J. Cloüet. *Note sur l'innocuité Absolue des Mélanges Colorants à Base de Fuchsine Pure.* Rouen, 1876.

Bericht über den XIV Internationale Kongress für Hygiene und Demographie Berlin, 23–29 September, 1907. Vol. II. Berlin, 1908.

"A Bill to Regulate the Importation, Manufacture and Sale of Butter Substitutes (Butter Substitutes Act 1887)." *Analyst* 12 (June 1887): 100.

"Birmingham Court Case." *British Food Journal & Analytical Review*, May 1900, 143.

"Blood Oranges." *Friendly Companion: A Magazine for Youth and the Home Circle*, November 1, 1890, 309.

Buchka, K. von. *Die Nahrungsmittelgesetzgebung im Deutschen Reiche.* Berlin: J. Springer, 1912.

Bujard, A., and E. Baier. *Hilfsbuch für Nahrungsmittelchemiker.* Berlin: J. Springer, 1894.

"By All That's Blue." *Fun*, July 26, 1873, 38.

Cameron, Charles. "Address given at the Meeting for the Society of Public Analysts, London, 1 August, 1885." *Analyst* 10 (October 1885): 175.

Cameron, Charles. "President's Address." *Analyst* 19 (March 1894): 55–57.

Caro, Heinrich. *Über die Entwicklung der Teerfarben-Industrie.* Friedländer, 1893.

Cassal, Charles. "On Dyed Sugar." *Analyst* 15 (1890): 141–60.

Cazeneuve, Paul. *Les Colorants de la Houille au Point de Vue Toxicologique et Hygiénique.* Lyon: Affaire de la succursale de la B. Anilin & Soda Fabrik à Neuvillesur-Saône, 1887.

Cazeneuve, P., and R. Lépine. "Les Couleurs de la Houille et la Revision des Listes Légales des Colorants Nuisibles et Non Nuisibles." *Annales d'hygiène* 17, 5 (1887): 6.

Cazeneuve, P., and R. Lépine. "Sur les Effets Produits par l'ingestion et l'infusion Intravéneuse de Trois Colorants, Dérivés de la Houille." *Comptes Rendus de Académie de Sciences* 101 (1885): 1167.

Charton, Edouard. *Guide pour le choix d'un état, ou, Dictionnaire des professions.* Paris: Librairie Vve Lenormant, 1842.

Chevallier, Alphonse. *Dictionnaire des Alterations et Falsifications.* Paris: Béchet, 1850.

"Chit-Chat." *Sheffield and Rotherham Independent* 11278 (1890): 5.

Chlopin, G. W. *Coal-Tar Dyes: Classification, Properties, and Action of Artificial Dyes on the Animal Organism.* Dorpat, 1903.

Clayton, Edwy Godwin. *Arthur Hill Hassall, Physician & Sanitary Reformer; a Short History of His Work in Public Hygiene, and of Movement Against the Adulteration of Food and Drugs.* London: Baillière, Tindall and Cox, 1908.

Cochin, Denys. "Les Falsificateurs et le Laboratoire Municipal." *Revue des Deux Mondes*, June 15, 1883, 861–87.

"Colour and Design in Ornamental Needlework." *Englishwoman's Domestic Magazine* 139 (July 1, 1876): 43.

The Country Gentleman: Sporting Gazette and Agricultural Journal 1076 (1882): 1322.

Curtman, C. O. "Test for Aniline Colours in Wines or Fruit Juices." *Zeitschrift für analytische Chemie* 4 (1887).

Detector. "Letter to the Editor." *Times* 31237 (September 12, 1884), 6.

Deutsche Chemische Gesellschaft. *Berichte der Deutschen Chemischen Gesellschaft*. Vol. 1. 1868.

Deutsches Nahrungsmittelbuch. Heidelberg: Carl Winters, 1905.

Dickens, Charles. *David Copperfield*. London, 1869.

Dupré, A. "On Copper in Food." *Analyst* 2, 13 (1877): 1b–4.

Dyer, B., C. A. Mitchell, Society of Public Analysts and Other Analytical Chemists (Great Britain), and Society of Public Analysts (Great Britain). *The Society of Public Analysts and Other Analytical Chemists: Some Reminiscences of Its First Fifty Years and a Review of Its Activities*. Cambridge: Heffer, 1932.

"Editorial." *Food, Drugs and Drink* 1, 24 (1893): 3.

"Editorial." *Food, Drugs and Drink* 2, 27 (1893): 1.

"Editorial." *Analyst* 2, 22 (1878): 171–72.

"Editorial and Exchange of Letters." *Analyst* 2, 24 (1878): 224.

Engelhardt, Roderich von. *Beiträge zur Toxikologie des Anilin*. Dorpat: H. Laakmann, 1888.

Erdmann, H. Review of Schultz and Julius, *Tabellarische Übersicht der künstlichen organischen Farbstoffe. Angewandte Chemie* 15, 30 (1902): 767.

Excise Office. *Report of the Principal Chemist*. London: H.M.S.O., 1860.

Feltz, V., and E. Ritter. *Etude expérimentale de l'action de la Fuscine sur l'organisme*. Paris: Nancy, 1877.

Filby, Frederick Arthur. *A History of Food Adulteration and Analysis*. London: G. Allen & Unwin, 1934.

Filehne, Dr Wilhelm. "Über die Giftwirkungen des Nitrobenzols." *Archiv für experimentelle Pathologie und Pharmakologie* 9, 5–6 (1878): 329–79.

"The First Bernard Dyer Memorial Lecture." *Analyst* 890, 75 (1950): 240.

Fletcher, Horace. *Fletcherism, What It Is: Or, How I Became Young at Sixty*. New York: Stokes, 1913.

"Food and Drugs (Adulteration) Bill." *British Medical Journal* 1 (March 19, 1898): 778.

Food and Sanitation, Formerly Food, Drugs and Drink, December 16, 1893, 387.

Food and Sanitation, November 28, 1896, 574.

"Forms of Adulteration." *Food, Drugs and Drink*, October 8, 1892, 5.

Fraenkel, Sigmund. *Die Arzneimittel Synthese auf Grundlage der Bezeitungen zwischen chemischen Aufbau und Wirkung*. Berlin: J. Springer, 1906.

Gautier, A. "Continuation of The Fraudulent Colouration of Wines." *Analyst* 1, 7 (1876): 130–35.

Gautier, A. "The Fraudulent Colouration of Wines." *Analyst* 1, 6 (1876): 109–12.

Gautier, A. "Ueber die Einwirkung des Chlorwasserstoffs." *Liebigs Annalen der Chemie*, 1867, 289.

Geisler, J. F. "A Delicate Test for the Detection of a Yellow Azo Dye Used for the Artificial Coloring of Fats." *Journal of the American Chemical Society* 20 (1898): 110–13.

Gibson, K. S. *The Lovibond Color System*. Washington, DC: United States Bureau of Standards, U.S. Government Printing Office, 1927.

Girard, Charles. *Documents sur les Falsifications des Matières Alimentaires et sur les Travaux du Laboratoire Municipal: Rapport à Monsieur le Préfet de Police: Deuxième Rapport*. Paris, 1885.

Goldstein, Max Aaron. *One Hundred Years of Medicine and Surgery in Missouri: Historical and Biographical Review of the Careers of the Physicians and Surgeons of the State of Missouri, and Sketches of Some of Its Notable Medical Institutions*. St. Louis: St. Louis Star, 1900.

"The Great Lozenge Maker." *Punch*, November 20, 1858.

Green, Arthur G., Gustav Schultz, and Paul Julius. *A Systematic Survey of the Organic Colouring Matters*. London: Macmillan, 1908.

"The Grocers' Federation: Report on Food Legislation." *British Food Journal and Analytical Review* 1 (February 1899): 45.

Gudeman, Edward. "Artificial Digestion Experiments." *Industrial and Engineering Chemistry* 27, 11 (1905): 1436–42.

"H. Kohnstamm & Co." *Oil, Paint and Drug Reporter* 101, 14 (March 1922): 122.

Halpen, G. "The Detection of Foreign Colouring Matters in Preserved Tomatoes." *Journal de Pharmacie et de Chimie* 11 (1900): 169–72.

Hassall, Arthur Hill. *Adulterations Detected, Or, Plain Instructions for the Discovery of Frauds in Food and Medicine*. London: Longman, Brown, Green, Longmans, and Roberts, 1857.

Hassall, Arthur Hill. *Food: Its Adulterations, and the Methods for Their Detection*. London: Longmans, Green, and Company, 1876.

Hassall, Arthur Hill, and Lancet Analytical Sanitary Commission. *Food and Its Adulterations: Comprising the Reports of the Analytical Sanitary Commission of "The Lancet" for the Years 1851 to 1854 Inclusive, Revised and Extended: Being Records of the Results of Some Thousands of Original Microscopical and Chemical Analyses of the Solids and Fluids Consumed by All Classes of the Public*. . . . London: Longman, Brown, Green, and Longmans, 1855.

Hasterlik, A. *Die praktische Lebensmittelkontrolle. Ein Leitfaden für die Nahrungs- und Genußmittelpolizei und für das Lebensmittelgewerbe*. Stuttgart: Eugen Ulmer, 1906.

"A Hat That Was Felt." *Funny Folks* 29 (June 26, 1875): 90.

Hehner, Otto. "Abstract of the Work of the Milk Committee." *Analyst* 11 (1886): 3–11.

Hehner, Otto. "On the Relation between the Specific Gravity, Fat and Solids-Non-Fat in Milk, upon the Basis of the Society of Public Analysts' Method." *Analyst* 13 (1882): 26–36.

Hehner, Otto. "On the Use of Food Preservatives." *Analyst* 15 (1890): 221–26.

Hehner, Otto. "President's Address: Public Analysts and the State." *Public Analytical Journal and Sanitary Review*, September 24, 1892, 8–9.

Herz, Joseph. "New Methods for Detecting Artificially Coloured Red Wines." *Chemiker Zeitung* 10 (1886).

Hesse, Bernhard C. *Coal-Tar Colors Used in Food Products*. Washington, DC: Government Printing Office, 1912.

Hesse, Bernard C. "The Industry of the Coal-Tar Dyes: An Outline Sketch." *Journal of Industrial & Engineering Chemistry* 6, 12 (1914): 1013–27.

Hofmann, August Wilhelm von. *Liebigs Annalen der Chemie*, 1867, 144–214.

Hofmann, August Wilhelm von. *The Life-Work of Liebig in Experimental and Philosophic Chemistry: With Allusions to His Influence on the Development of the Collateral Sciences and of the Useful Arts; a Discourse Delivered to the Fellows of the Chemical Society of London in the Theatre of the Royal Institution of Great Britain, on March the 18th, 1875*. London: Macmillan, 1876.

Hofmann, August Wilhelm von. *On Mauve and Magenta: A Lecture, Delivered on Friday, April 11, 1862, in the Theatre of the Royal Institution of Great Britain*. London, 1862.

Hofmann, August Wilhelm von. *On the Importance of the Study of Chemistry: Delivered at The South Kensington Museum, 7th January 1861*. London, 1868.

"How We Are Poisoned in 1890." *Illustrated Chips* 6 (August 30, 1890): 92.

Hueppe, Ferdinand. *Die Methoden der Bakterien-Forschung*. Wiesbaden: C. W. Kreidel, 1885.

Huet-Desaunay, Henry. *Le Laboratoire Municipal et les Falsifications ou Recueil des Lois et Circulaires Concernant la Vente des Produits Alimentaires et Hygiene Publique*. Paris, 1890.

International Exhibition, 1862: Reports by the Juries on the Subjects in the Thirty-Six Classes Into Which the Exhibition Was Divided. London: Society of Arts, 1863.

International Labour Office. *Cancer of the Bladder among Workers in Aniline Factories*. Geneva: International Labour Office, 1921.

"International Legislation on Adulteration of Food." *British Medical Journal* 2, 1291, 1885, 607–8.

Kayser, R. "Ueber die Beurtheilung von Farbstoffen hinsichtlich ihrer Gesundheits-schädlichkeit." *Forschungsberichte über Lebensmittel* 2 (1895): 181.

Kellogg, John Harvey. *The New Dietetics: A Guide to Scientific Feeding in Health and Disease*. Battle Creek, MI: Modern Medicine Publishing Co, 1927.

Kerp, W. *Nahrungsmittelchemische Tagesfragen: Über die durch die gewerbliche Herstellung der Lebensmittel an diesen hervorgebrachten Erscheinungen*. Berlin: J. Springer, 1914.

Kingsley, Charles. *The Water-Babies (a Fairy Tale for a Land-Baby)*. Edited by Warwick Goble. London: Macmillan, 1863.

Kohnstamm, H., & Co. "The Development of Certified Pure Food Colors." In

Chemical Industry's Contribution to the Nation: 1635–1935. ed. William Haynes and Edward L. Gordy. New York: Chemical Markets, 1935: 5–16.

König, J. *Chemie der menschlichen Nahrungs- und Genussmittel.* Vol. 2. Berlin: J. Springer, 1904.

König, J., and A. Juckenack. "Preußen." In *Die Anstalten zur technischen Untersuchung von Nahrungs- und Genußmitteln sowie Gebrauchsgegenständen, die im Deutschen Reiche,* edited by Dr J. König and Dr A. Juckenack, 3–159. Berlin, Heidelberg: Springer, 1907.

Leeds, Albert R. "Methods for Separating Out Colours in Butter, Imitation Butter," *Analyst* 12 (September 1887): 150–51.

Lehmann, Karl. *Methoden der praktischen Hygiene.* Wiesbaden: Bergmann, 1890.

Lepsius, B. *Festschrift zur Feier des 50 jährigen Bestehens der Deutschen Chemischen Gesellschaft und des 100. Geburtstages ihres Begründers August Wilhelm von Hofmann.* Berlin: Friedländer, 1918.

"The Letter Box." *St. Nicholas* 6 (April 1, 1875): 388.

Lieber, Hugo. *The Use of Coal Tar Colors in Food Products.* New York: Bergmann, 1904.

Liebig, Justus. *Researches on the Chemistry of Food.* London: Taylor and Walton, 1847.

Linnaeus, Carl, William Thomas Stearn, and J. L. Heller. *Species Plantarum: Carl Linnaeus.* London: Bartholomew, 1753.

Lovibond, Joseph W. "On the Scientific Measurement of Colour in Beer." *Journal of the Federated Institutes of Brewing* 3, 5 (1897): 405–29.

Mackay, Thomas. *A Plea for Liberty.* New York: D. Appleton and Company, 1891.

Martin, Edward G. "Detection of Artificial Colouring Matter in Butter, Oleomargarine, Fats, Oils Etc." *Analyst* 12 (April 1887): 70.

Martin, Edward G. "Methods of Separating and Determining Artificial Colours in Butter." *Analyst* 10 (September 1885): 161.

Martius, C. A. von. *Chemische Erinnerungen aus der Berliner Vergangenheit: zwei akademische Vorträge.* Berlin: Hirschwald, 1918.

"Meat-Tints for the Million." *Funny Folks* 174 (1878): 101.

"Meeting of Public Analysts." *Chemical News,* August 14, 1874, 73.

Micko, K. "Artificial Colouring of Oranges." *Zeitschrift für Untersuchung der Nahrungs- und Genussmittel* 3 (1900): 729–35.

Munsell, A. H. (Albert Henry). *A Color Notation: A Measured Color System, Based on the Three Qualities Hue, Value and Chroma.* New York: G. H. Ellis, 1907.

Munsell, A. H. "A Pigment Color System and Notation." *American Journal of Psychology* 23, 2 (1912): 236–44.

Muter, John. "On the Processes and Standards in Use at the Municipal Laboratory of the City of Paris." *Analyst* 10 (October 1885): 179–81.

Muter, John. "On the Processes and Standards of Food Analysis in Use at the Municipal Laboratory of the City of Paris." *Analyst* 10 (August 1985): 143–45.

Muter, John. "President's Address." *Analyst* 5 (February 1880): 15–16.

"Mythology and Socks." *Punch*, October 7, 1868, 160.

Natrajan, T. S., Bernard Dyer, and G. C. Clayton. "Obituary Notices: George Herbert Bailey, 1852–1924; Surendra Nath Dhar, 1892–1923; James Johnston Dobbie, 1852–1924; Otto Hehner, 1853–1924; Edmund Knowles Muspratt, 1833–1923." *Journal of the Chemical Society, Transactions* 125 (January 1, 1924): 2677–98.

Neufeld, C. A. *Der Nahrungsmittelchemiker als Sachverständiger: Anleitung zur Begutachtung der Nahrungsmittel, Genussmittel und Gebrauchsgegenstände nach den gesetzlichen Bestimmung, mit praktischen Beispielen.* Berlin: J. Springer, 1907.

Nordau, Max Simon. *Degeneration.* New York: D. Appleton, 1895.

"Notes of the Month." *Analyst* 2, 23 (1878): 205–6.

"Notes of the Month." *Analyst* 3, 29 (1878): 316.

Official Catalogue of the Great Exhibition of the Works of Industry of All Nations, 1851. Unknown, 2010.

Oleomargarine and Butterine: A Plain Presentation of the Most Gigantic Swindle of Modern Times. New York: T. L. McAlpine, 1886.

"Organization Amongst Chemists." *Analyst* 2, 19 (1877): 109–11.

O'Shaughnessy, W. B. *Poisoned Confectionary.* London: Mills, Jowett, and Mills, 1831.

Perkin, William Henry. *Perkin Centenary London: 100 Years of Synthetic Dyestuffs.* London: Pergamon, 1958.

"Perkins's [sic] Purple." *All the Year Round.* September 1859, 222.

Piesse, Charles H. "Copper in Preserved Green Peas." *Analyst* 2, 14 (1877): 27.

"Poisoned Candies." *Health Reformer* 6–7, 1871, 131.

"The Poisoned Hat." *Punch*, June 19, 1875, 262.

"Poisoning by Coloured Silk Stockings." *Englishwoman's Domestic Magazine* 164 (August 1, 1878): 109.

"Poisonous Ice Cream." *Analyst* 3, 29 (1878): 311.

"Proceedings of the Society of the Public Analysts." *Analyst* supplement (April 1894): 110.

"Progress and Butter." *Puck*, March 1, 1880, 55.

"The Public and 'Public Analysts.'" *Analyst* 1, 9 (1876): 155–56.

"Pure Food." *British Medical Journal*, December 1, 1896, 1794.

"Queer Street." *Review of Reviews* 14 (1896): 79.

"A Ramble into the Eastern Annexe of the International Exhibition." *Ladies Treasury*, November 1, 1862, 342.

Rassow, Bertold. *Geschichte des Vereins Deutsche Chemiker in dem ersten 25 Jahren.* Leipzig, 1912.

Redlich, Fritz. "Die volkswirtschaftliche Bedeutung der deutschen Teerfarbenindustrie." Doctoral dissertation, Friedrich-Wilhelms-Universität, 1914.

"Remuneration of Public Analysts." *Nature*, March 14, 1912, 34.

"Report of Meeting." *Analyst* 2, 23 (1878): 189–90.

"Report of the Annual Meeting." *Analyst* 4, 35 (1879): 21.

Richards, Edgar. "Certain Provisions of Continental Legislation Concerning Food Adulteration." *Science* 14, 353 (1889): 308–10.

Richards, Edgar. "Legislation on Food Adulteration." *Science* 16, 394 (1890): 101–4.

Rideal, Samuel. "An Investigation of Certain Substances Used in Colouring Foods." *Lancet* 177, 4580 (1911): 1597–1601.

Rinzand, M. "Artificial Colouration of Wine." *Union Pharmaceutique* 36 (1895): 446.

Rost, E. *Pharmakologische Untersuchung des Eosins mit Berücksichtigung der Wirknungen des Fluoreszeins und Erythrosins.* Berlin: Springer, 1912.

Rota, A. R. "A Method of Analyzing Natural and Artificial Organic Colouring Matters." *Chemiker Zeitung* 1898: 437–42.

Rowe, F. M., ed. "Colour Index. Part I. Bradford: The Society of Dyers and Colourists, 1922. Subscription Price of the Whole Work, in Monthly Parts, 84s." *Journal of the Society of Chemical Industry* 41, 22 (November 30, 1922).

Runge, F. F. "Ueber einige Produkte der Steinkohlendestillation." *Annalen der Physik und Chemie* 31 (1844): 65–77, 513–24; 32 (1844): 308–32.

Rupke, Nicholaas. *Vivisection in Historical Perspective.* London: Croom Helm, 1887.

Rupp, J. G. *Die Untersuchung von Nahrungsmitteln, Genussmitteln und Gebrauchsgegenständen. Praktisches Handbuch für Chemiker.* Heidelberg: C. Winter, 1900.

"Scene: Commercial Room." *Punch*, November 23, 1861, 212.

Schlacherl, Gustav. *Fifth International Congress of Applied Chemistry, Berlin* 3 (1903): 1041–48.

Schödler, Friedrich. "Das chemische Laboratorium unserer Zeit." *Westermann's Monatshefte* 38 (1875): 21–27.

Schuchardt, B. "Ueber die Wirkungen des Anilins auf den thierischen Organismus." *Archiv der Pharmazie* 156, 2 (1861): 144–64.

Schultz, G., and P. Julius. Translated and edited, with extensive additions, by Arthur G. Green. *Systematic Survey of the Organic Colouring Matters.* London: Macmillan and Co., 1908.

Schultz, G., and P. Julius. *Tabellarische Übersicht der künstlichen organischen Farbstoffe.* Berlin: R. Gaertners Verlag, 1902.

Schweissinger, Otto. "Microscopic Detection of Colouring Matters in Sausages." *Analyst* 12 (March 12, 1887): 53.

Schweitzer, H. "Ehrlich's Chemotherapy: A New Science." *Science*, n.s., 32, 832 (1910): 809–23.

"Scientific Facts." *Ladies' Treasury*, April 1, 1861, 116.

Scott Elder, B. *Appeal Cases Under the Sale of Food & Drugs Acts, 1875 & 1879, and the Margarine Act, 1887.* London: Butterworth, 1900.

Snodgrass, Katharine. *Margarine as a Butter Substitute.* Stanford, CA: Food Research Institute, 1930.

"Some Very Ancient Things." *Ladies' Treasury*, August 1, 1865, 242.

"The Somerset House Court of Appeal." *Analyst* 2, 20 (1877): 2, 23.

Spaeth, E. "The Detection of Artificial Colouring Matters in Sausages." *Zeitschrift für Untersuchung der Nahrungs- und Genussmittel* 4 (1901): 1020–23.

Spaeth, E. "On Fruit Juices and Their Examination, with Particular Reference to Raspberry Juice: Detection of Foreign Colouring Matter." *Zeitschrift für Untersuchung der Nahrungs- und Genussmittel* 2 (1899): 633–35.

Spencer, H. "From Freedom to Bondage." Introduction in Thomas Mackay, *A Plea for Liberty*. New York: D. Appleton and Company, 1891.

Spiekermann, Uwe. "Twentieth-Century Product Innovations in the German Food Industry." *Business History Review* 83, 2 (2009): 291–315.

Stilling, Jakob. *Anilinfarbstoffe als Antiseptica*. Vol. 2. Strassburg: Trübner, 1890.

Taine, Hippolyte. *Taine's Notes on England*. Translated by Hyams Edward. London: Thames and Hudson, 1957.

"Talk about Depression in Trade." *Moonshine*, August 7, 1886, 72.

"Talk with Our Readers." *Englishwoman's Domestic Magazine* 187, 131 (1875): 272.

"The Thames and Its Deodorization." *John Bull & Britannia*, 2074, 1860, 569.

Thorpe, T. E. "Obituary of Gautier." *Nature* 106 (September 16, 1920): 85–86.

Turnbull, James. "On the Physiological and Medicinal Properties of Sulphate of Aniline, and Its Use in the Treatment of Cholera." *Lancet* 78, 1994 (1861): 469–71.

"Useful Book." *Ladies' Treasury*, January 1, 1875, 42.

Vereinbarungen zur einheitlichen Untersuchung und Beurtheilung von Nahrungs- und Genußmitteln sowie Gebrauchständen für das Deutsche Reich: Ein Entwurf festgestellt nach den Beschlüssen der auf Anregung des Kaiserlichen Gesundheitamtes einberufenen Kommission deutscher Nahrungsmittelchemiker. Berlin: J. Springer, 1897–1902.

"Vorsätze der Schweizer Analytiker." *Zeitschrift für Ernährungswissenschaft* 5 (1891): 293.

Warnford Lock, Charles G., Benjamin E. R. Newlands, and John A. R. Newlands. *Sugar: A Handbook for Planters and Refiners*. London: F. N. Spon, 1888.

Weber, H. A. "On the Behaviour of Coal-Tar Colours toward the Process of Digestion." *American Chemical Journal* 18, 12 (1896): 1092–96.

Weyl, Theodor. *Handbuch der Hygiene*. Vol. 3. Jena: Gustav Fischer, 1896.

Weyl, Theodor. "Ueber eine neue Reaction auf Kreatinin und Kreatin." *Berichte der Deutschen Chemischen Gesellschaft* 11, 2 (July 1, 1878): 2175–77.

Weyl, Theodor, and Henry Leffman. *The Coal-Tar Colors: With Especial Reference to Their Injurious Qualities and the Restriction of Their Use*. Philadelphia: P. Blakiston, 1892.

"The Whole Duty of a Chemist," *Nature* 33 (November 16, 1885): 73–77.

Wiley, Harvey Washington. *The History of a Crime against the Food Law; the Amazing Story of the National Food and Drugs Law Intended to Protect the Health of the People, Perverted to Protect Adulteration of Foods and Drugs*. Washington, DC: Wiley, 1929.

Wiley, Harvey W., and Anne Lewis Pierce, *1001 Tests of Foods, Beverages and Toilet Accessories, Goods and Otherwise; Why They Are So*. New York: Hearst, 1914.

[Williams, John Dingwall.] *Deadly Adulteration and Slow Poisoning: Or, Dis-*

ease and Death in the Pot and the Bottle; in Which the Blood-Empoisoning and Life-Destroying Adulterations of Wines, Spirits, Beer, Bread, Flour, Tea, Sugar, Spices, Cheesemongery, Pastry, Confectionary Medicines, &c. &c. &c. Are Laid Open to the Public, with Tests or Methods for the Ascertaining and Detecting the Fraudulent and Deleterious Adulterations and the Good and Bad Qualities of Those Articles: With an Exposé of Medical Empiricism and Imposture, Quacks and Quackery, Regular and Irregular, Legitimate and Illegitimate: And the Frauds and Mal-Practices of Pawn-Brokers and Madhouse Keepers. London: Sherwood, Gilbert and Piper, 1830.

Winton, A. L. *Connecticut Agricultural Experiment Station Report.* Connecticut State Stationery Office, 1901.

SECONDARY SOURCES

Abelshauser, Werner, Wolfgang von Hippel, Jeffrey Allan Johnson, and Raymond G. Stokes. *German Industry and Global Enterprise: BASF: The History of a Company.* Cambridge: Cambridge University Press, 2003.

Adam, Barbara, Ulrich Beck, and Joost Van Loon, eds. *The Risk Society and Beyond: Critical Issues for Social Theory.* Thousand Oaks, CA: Sage Publications Ltd., 2000.

Albala, Ken. Keynote address. Presented at the Anglo-American Conference of Historians, London, July 11, 2013.

Allen, Michelle Elisabeth. *Cleansing the City: Sanitary Geographies in Victorian London.* Athens: Ohio University Press, 2008.

Altick, Richard D. *The English Common Reader: A Social History of the Mass Reading Public: 1800–1900.* Chicago: University of Chicago Press, 1983.

Anderson, Margaret Jean. *Carl Linnaeus: Father of Classification.* New York: Enslow Publishers, Inc., 2009.

Anderson, Oscar E. *The Health of a Nation: Harvey W. Wiley and the Fight for Pure Food.* Chicago: University of Chicago Press, 1958.

Arnold, L. Eugene, Nicholas Lofthouse, Elizabeth Hurt. "Artificial Food Colors and Attention-Deficit/Hyperactivity Symptoms: Conclusions to Dye For." *Neurotherapeutics* 9, 3 (2012): 599–609.

Ash, Eric, ed. "Expertise and the Early Modern State." Special issue, *Osiris* 25 (2010).

Ash, Mitchell G., and Jan Surman. *The Nationalization of Scientific Knowledge in the Habsburg Empire, 1848–1918.* Basingstoke, Hampshire: Palgrave Macmillan, 2012.

Ashworth, William J. *Customs and Excise: Trade, Production, and Consumption in England, 1640–1845.* Oxford: Oxford University Press, 2003.

Atkins, Peter. *Liquid Materialities: A History of Milk, Science, and the Law.* Farnham: Ashgate Publishing, Ltd., 2010.

Atkins, Peter. "The Material Histories of Food Quality and Composition." *Endeavour* 35, 2–3 (June 2011): 74–79.

Atkins, Peter. "Sophistication Detected: Or, the Adulteration of the Milk Supply, 1850–1914." *Social History* 16, 3 (1991): 317–39.

Atkins, Peter. "Vinegar and Sugar: The Early History of Factory-Made Jams,

Pickles, and Sauces in Britain." In *The Food Industries of Europe in the Nineteenth and Twentieth Centuries*, edited by Derek J. Oddy and Alain Drouard, 41–55. Farnham: Ashgate, 2013.

Atkins, Peter, Peter Lummel, and Derek J. Oddy, eds. *Food and the City in Europe since 1800*. Farnham: Ashgate, 2007.

Atkins, Peter, and A. Stanziani. "From Laboratory Expertise to Litigation: The Municipal Laboratory of Paris and the Inland Revenue Laboratory in London, 1870–1914." In *Fields of Expertise: A Comparative History of Expert Procedures in Paris and London, 1600 to Present*, edited by Christelle Rabier, 317–39. Cambridge: Cambridge Scholars Publishing, 2007.

Auerbach, Jeffrey. *The Great Exhibition of 1851: A Nation on Display*. New Haven, CT: Yale University Press, 1999.

Ball, Philip. *Bright Earth: Art and the Invention of Color*. Chicago: University of Chicago Press, 2003.

Ball, Richard A., and J. Robert Lilly. "The Menace of Margarine: The Rise and Fall of a Social Problem." *Social Problems* 29, 5 (1982): 488–98.

Barker, Theodore Cardwell, John Crawford Mackenzie, and John Yudkin. *Our Changing Fare: Two Hundred Years of British Food Habits*. London: Macgibbon & Kee, 1966.

Bashford, A. *Imperial Hygiene: A Critical History of Colonialism, Nationalism, and Public Health*. Basingstoke: Palgrave Macmillan, 2004 [2003].

Baudrillard, Jean. *The Consumer Society: Myths and Structures*. Thousand Oaks, CA: Sage Publications Ltd., 1998.

Beck, Ulrich. "Critical Theory of World Risk Society: A Cosmopolitan Vision." *Constellations* 16, 1 (2009): 3–22.

Beck, Ulrich. *Risk Society: Towards a New Modernity*. Thousand Oaks, CA: Sage Publications Ltd., 1992.

Beer, Gillian. *Open Fields: Science in Cultural Encounter*. Oxford: Oxford University Press, 1999.

Beer, John J. "Coal Tar Dye Manufacture and the Origins of the Modern Industrial Research Laboratory." *Isis* 49, 2 (1958): 123–31.

Beer, John J. *The Emergence of the German Dye Industry*. Chicago: University of Illinois Press, 1959.

Beetham, Margaret, and Kay Boardman. *Victorian Women's Magazines: An Anthology*. Manchester: Manchester University Press, 2001.

Beguet, Bruno, ed. *La Science pour tous: La Vulgarisation Scientifique en France de 1850 à 1914*. Paris: Bibliotheque du Conservatoire National des Arts et Metiers, 1990.

Belt, Henk van den, and Arie Rip. "The Nelson-Winter-Dosi Model and Synthetic Dye Chemistry." In *The Social Construction of Technological Systems: New Directions in the Sociology and History of Technology*, edited by Wiebe E. Bijker, Thomas P. Hughes, and Trevor Pinch, 129–54. Cambridge MA: MIT Press, 2012.

Ben-David, Joseph. *Centers of Learning: Britain, France, Germany, United States*. New Brunswick, NJ: Transaction Publishers, 2008.

Ben-David, Joseph. *The Scientist's Role in Society, a Comparative Study*. Englewood Cliffs, NJ: Prentice Hall, 1971.

Bender, David A. *A Dictionary of Food and Nutrition*. Oxford: Oxford University Press, 2009.

Bensaude-Vincent, Bernadette. *A History of Chemistry*. Cambridge, MA: Harvard University Press, 1996.

Bensaude-Vincent, Bernadette, and Ferdinando Abbri. *Lavoisier in European Context: Negotiating a New Language for Chemistry*. London: Science History Publications, 1995.

Bensaude-Vincent, Bernadette, and Jonathan Simon. *Chemistry: The Impure Science*. London: Imperial College Press, 2008.

Benson, John. *The Rise of Consumer Society in Britain, 1880–1980*. London: Longman, 1994.

Berenstein, Nadia. *Flavor Added* (blog). www.nadiaberenstein.com/blog.

Berenstein, Nadia. *The Inexorable Rise of Synthetic Flavor: A Pictorial History*. www.popsci.com, November 23, 2015.

Berghahn, Volker Rolf. *Imperial Germany, 1871–1918: Economy, Society, Culture, and Politics*. Oxford: Berghahn Books, 2005.

Berman, Morris. *Social Change and Scientific Organization: The Royal Institution, 1799–1844*. London: Heinemann Educational, 1978.

Bernhardt, Emily S., Emma J. Rosi, and Mark O. Gessner. "Synthetic Chemicals as Agents of Global Change." *Frontiers in Ecology and the Environment* 15, 2 (2017): 84–90.

Bijker, W., Thomas Hughes, and Trevor Pinch. *The Social Construction of Technological Systems: New Directions in the Sociology and History of Technology*. New ed. Cambridge, MA: MIT Press, 1989.

Blaszczyk, Regina Lee. *The Color Revolution*. Cambridge, MA: MIT Press, 2012.

Blaszczyk, Regina Lee, and Uwe Spiekermann, eds. *Bright Modernity: Colour, Commerce, and Consumer Culture*. Cham, Switzerland: Palgrave Macmillan, 2017.

Blunt, Wilfrid. *Linnaeus: The Compleat Naturalist*. London: Frances Lincoln, 2004.

Blythman, Joanna. *Swallow This*. London: Harper Collins, 2015.

Boas, Marie. *Robert Boyle and Seventeenth-Century Chemistry*. Cambridge: Cambridge University Press Archive, 1958.

Boddice, Rob. "Species of Compassion: Aesthetics, Anaesthetics, and Pain in the Physiological Laboratory." *Interdisciplinary Studies in the Long Nineteenth Century* 19, 15 (2012).

Bodewitz, H., H. Buurma, and G. H. de Vries. "Regulatory Science and the Social Management of Trust in Medicine." In *The Social Construction of Technological Systems: New Directions in the Sociology and History of Technology*, edited by Wiebe E. Bijker, Thomas P. Hughes, and Trevor Pinch, 237–52. Cambridge, MA: MIT Press, 2012.

Bogousslavsky, Julien. *Following Charcot: A Forgotten History of Neurology and Psychiatry*. Basel: Karger Medical and Scientific Publishers, 2011.

Boisard, Pierre. *Camembert: A National Myth*. Berkeley: University of California Press, 2003.

Boswell, James, ed. *JS 100: The Story of Sainsbury's*. London: J. Sainsbury Ltd, 1969.

Boudia, Soraya. "Global Regulation: Controlling and Accepting Radioactivity Risks." *History and Technology* 23, 4 (2007): 389–406.

Boudia, Soraya, Angela Creager, Scott Frickel, Emmanuel Henry, Nathalie Jas, Carsten Reinhardt, and Jody A. Roberts. "Residues: Rethinking Chemical Environments." *Engaging Science, Technology, and Society* 4 (2018).

Boudia, Soraya, and Nathalie Jas. "Introduction: Risk and 'Risk Society' in Historical Perspective." *History and Technology* 23, 4 (2007): 317–31.

Boudia, Soraya, and Nathalie Jus. *Toxicants, Health, and Regulation since 1945* Abingdon, UK: Taylor & Francis, 2013.

Bourdieu, Jérôme, Martin Bruegel, and Peter Atkins. "'That Elusive Feature of Food Consumption: Historical Perspectives on Food Quality, a Review and Some Proposals." *Food & History* 5, 2 (2007): 247–66.

Bourdieu, Pierre. *Distinction: A Social Critique of the Judgement of Taste.* Cambridge, MA: Harvard University Press, 1984.

Brears, Peter. *Jellies and Their Moulds.* Totnes: Prospect Books, 2010.

Brewer, John. *Consumption and the World of Goods.* London: Taylor & Francis Group, 1994.

Brimblecombe, Peter. *The Big Smoke.* Abingdon: Routledge, 2012.

Brock, W. H. "Breeding Chemists." *Ambix* 50, 1 (2003): 25–70.

Brock, W. H. *The Case of the Poisonous Socks: Tales from Chemistry.* London: Royal Society of Chemistry, 2011.

Brock, W. H. *Justus von Liebig: The Chemical Gatekeeper.* Cambridge: Cambridge University Press, 1997.

Brock, W. H. *William Crookes (1832–1919) and the Commercialization of Science.* Farnham: Ashgate Publishing, Ltd., 2008.

Broks, Peter. *Media Science Before the Great War.* Basingstoke: Palgrave Macmillan, 1996.

Broomfield, Andrea. "Rushing Dinner to the Table: The Englishwoman's Domestic Magazine and Industrialization's Effects on Middle-Class Food and Cooking, 1852–1860." *Victorian Periodicals Review* 41, 2 (2008): 101–23.

Brown, Lucy. *Victorian News and Newspapers.* Oxford: Oxford University Press, 1985.

Brunello, Franco. *The Art of Dyeing in the History of Mankind.* Vicenza: AATCC, 1973.

Buchwald, Jed Z. *Scientific Credibility and Technical Standards in 19th and Early 20th Century Germany and Britain.* Berlin: Springer Science & Business Media, 1996.

Bud, Robert, and Gerrylynn K. Roberts. *Science Versus Practice: Chemistry in Victorian Britain.* Manchester: Manchester University Press, 1984.

Bud, Robert, and Deborah Jean Warner. *Instruments of Science: An Historical Encyclopedia.* London: Taylor & Francis, 1998.

Burchardt, Lothar. "Professionalisierung oder Berufskonstruktion?: das Beispiel des Chemikers im wilhelminischen Deutschland." *Geschichte und Gesellschaft* 6, 3 (1980): 326–48.

Burchardt, Lothar. "Die Zusammenarbeit zwischen chemischer Industrie, Hochschulchemie und chemischen Verbänden in Wilhelmischen Deutschland." *Technikgeschichte* 46 (1979): 194.

Burdett, B. C. "The Colour Index: The Past, Present, and Future of Colorant Classification." *Journal of the Society of Dyers and Colourists* 98, 4 (1982): 114–20.

Burnett, John. *Plenty and Want: A Social History of Food in England from 1815 to the Present Day*. Abingdon: Routledge, 1989.

Burney, Ian A. *Bodies of Evidence: Medicine and the Politics of the English Inquest, 1830–1926*. Baltimore: Johns Hopkins University Press, 2000.

Burney, Ian A. *Poison, Detection, and the Victorian Imagination*. Manchester: Manchester University Press, 2006.

Burns, Christopher. "Bogus Butter: An Analysis of the 1886 Congressional Debates on Oleomargarine Legislation." Ph.D. diss., University of Vermont, 2009.

Burns, D. Thorburn. "Analytical Chemistry and the Law: Progress for Half a Millennium." *Fresenius' Journal of Analytical Chemistry* 368, 6 (2000): 544–47.

Burns, D. Thorburn. "Sir Charles Cameron (1830–1931): Dublin's Medical Superintendant, Executive Officer of Public Health, Public Analyst and Inspector of Explosives." *Journal of the Association of Public Analysts* 37 (2009): 14–39.

Burns, Edward. *Bad Whisky: The Scandal That Created the World's Most Successful Spirit*. Castle Douglas, Scotland: Neil Wilson Publishing, 2011.

Burrows, Adam. "Palette of Our Palates: A Brief History of Food Coloring and Its Regulation." *Comprehensive Reviews in Food Science and Food Safety* 8, 4 (2009): 394–408.

Cahan, David. *From Natural Philosophy to the Sciences: Writing the History of Nineteenth-Century Science*. Chicago: University of Chicago Press, 2003.

Caivano, José Luis, and María del Pilar Buera. *Color in Food: Technological and Psychophysical Aspects*. Boca Raton, FL: CRC Press, 2012.

Callon, Michel, Pierre Lascoumes, and Yannick Barthe. *Acting in an Uncertain World: An Essay on Technical Democracy*. Cambridge, MA: MIT Press, 2011.

Campbell, W. A. *Chemical Industry*. London: Prentice Hall Press, 1971.

Cantor, Geoffrey, Gowan Dawson, Graeme Gooday, Richard Noakes, Sally Shuttleworth, and Jonathan R. Topham. *Science in the Nineteenth-Century Periodical: Reading the Magazine of Nature*. Cambridge: Cambridge University Press, 2004.

Carpenter, Kenneth J. *Protein and Energy: A Study of Changing Ideas in Nutrition*. Cambridge: Cambridge University Press, 1994.

Carson, Rachel. *Silent Spring*. Mariner Books, 1962.

Cartwright, Nancy. *Nature's Capacities and Their Measurements*. Oxford: Clarendon Press, 1994.

Case, R. A., and J. T. Pearson. "Tumours of the Urinary Bladder in Workmen Engaged in the Manufacture and Use of Certain Dyestuff Intermediates in the British Chemical Industry: Further Consideration of the Role of Aniline and of the Manufacture of Auramine and Magenta (Fuchsine) as Possible Causative Agents." *British Journal of Industrial Medicine* 11, 3 (1954): 213–16.

Chadwick, Edwin. "The British Sanitary Movement." In *Eras in Epidemiology*, edited by Mervyn Susser and Zena Stein, 50–65. Oxford: Oxford University Press, 2009.

Chang, Hasok. "Circularity and Reliability in Measurement." *Perspectives on Science* 3 (1995): 153–72.

Chang, Hasok. *Inventing Temperature: Measurement and Scientific Progress*. Oxford: Oxford University Press, 2007.

Chang, Hasok. *Is Water H_2O?: Evidence, Realism, and Pluralism*. Dordrecht, Netherlands: Springer, 2012.

Chang, Hasok, and Nancy Cartwright. "Measurement." In *The Routledge Companion to Philosophy of Science*, edited by Martin Curd and Stathis Psillos. London: Routledge, 2008.

Chast, François. "Les Colorants, Outils Indispensables de la Révolution Biologique et Thérapeutique du XIXe Siècle." *Revue d'histoire de la Pharmacie* 93, 348 (2005): 487–504.

Chatriot, Alain, Marie-Emmanuelle Chessel, and Matthew Hilton. *The Expert Consumer: Associations and Professionals in Consumer Society*. Aldershot: Ashgate Publishing, Ltd., 2006.

Chirnside, R. C., and J. H. Hamence. *"Practising Chemists": History of the Society for Analytical Chemistry, 1874–1974*. London: Royal Society of Chemistry, 1974.

Civitello, Linda. *Baking Powder Wars: The Cutthroat Food Fight that Revolutionized Cooking*. Chicago: University of Illinois Press, 2017.

Clark, George, Frederick H. Kasten, and Harold Joel Conn. *History of Staining*. Philadelphia: Williams & Wilkins, 1983.

Clark, M. *Handbook of Textile and Industrial Dyeing: Principles, Processes, and Types of Dyes*. London: Elsevier, 2011.

Cleland, T. M. *A Practical Description of the Munsell Color System and Suggestions for Its Use 1937*. Whitefish, MT: Kessinger Publishing, 2010.

Clydesdale, Fergus M. "Color as a Factor in Food Choice." *Critical Reviews in Food Science and Nutrition* 33, 1 (1993): 83–101.

Cobbold, Carolyn. "The Rise of Alternative Bread Leavening Technologies in the Nineteenth Century." *Annals of Science* 75, 1 (2018): 21–39.

Cochrane, Rexmond Canning. *Measures for Progress: A History of the National Bureau of Standards*. Washington, DC: National Bureau of Standards, U.S. Department of Commerce, 1966.

Cocks, Geoffrey, and Konrad Hugo Jarausch. *German Professions, 1800–1950*. Oxford: Oxford University Press, 1990.

Coff, Christian. *The Taste for Ethics: An Ethic of Food Consumption*. Berlin: Springer, 2006.

Cohen, Benjamin R. *Pure Adulteration: Cheating on Nature in the Age of Manufactured Food*. Chicago: University of Chicago Press, 2019.

Collard, Patrick John, and Patrick Collard. *The Development of Microbiology*. Cambridge: CUP Archive, 1976.

Collins, E. J. T. "Food Adulteration and Food Safety in Britain in the 19th and Early 20th Centuries." *Food Policy* 18, 2 (1993): 95–109.

Collins, Edward, and Derek J. Oddy, "The Centenary of the British Food Journal, 1899–1999: Changing Issues in Food Safety Regulation and Nutrition." *British Food Journal* 100, 10/11 (1998): 433–550.

Collins, H. M. *Changing Order: Replication and Induction in Scientific Practice.* Chicago: University of Chicago Press, 1992.

Collins, H. M. "Son of Seven Sexes: The Social Destruction of a Physical Phenomenon." *Social Studies of Science* 11, 1 (1981): 33–62.

Collins, H. M., and Robert Evans. *Rethinking Expertise.* Chicago: University of Chicago Press, 2008.

Collins, H. M., and Robert Evans. "The Third Wave of Science Studies: Studies of Expertise and Experience." *Social Studies of Science* 32, 2 (2002): 235–96.

Conti, A., and M. Bickel. "History of Drug Metabolism: Discoveries of the Major Pathways of the Nineteenth Century." *Drug Metabolism Reviews* 6, 1 (1977): 1–50.

Cook, H. C. "Origins of Tinctorial Methods in Histology." *Journal of Clinical Pathology* 50, 9 (September 1997): 716–20.

Cooter, R., and S. Pumfrey. "Separate Spheres and Public Places: Reflections on the History of Science Popularization and Science in Popular Culture." *History of Science* 32 (1994): 237–67.

Coppin, Clayton A., and Jack C. High. *The Politics of Purity: Harvey Washington Wiley and the Origins of Federal Food Policy.* Ann Arbor: University of Michigan Press, 1999.

Corley, T. A. B. *Huntley and Palmers of Reading: A Business History.* London: Hutchison, 1972.

Coveney, John. *Food, Morals, and Meaning: The Pleasure and Anxiety of Eating.* Abingdon: Routledge, 2002.

Crosland, Maurice. "The Organisation of Chemistry in Nineteenth-Century France." In *The Making of the Chemist: The Social History of Chemistry in Europe, 1799–1914,* edited by David M. Knight and Helge Kragh. Cambridge: Cambridge University Press, 1998.

Curd, Martin, and Stathis Psillos, eds. *The Routledge Companion to Philosophy of Science.* London: Routledge, 2008.

Daunton, Martin, and Matthew Hilton. *The Politics of Consumption: Material Culture and Citizenship in Europe and America.* Oxford: Berg, 2001.

Davenport, Horace W. *A History of Gastric Secretion and Digestion: Experimental Studies to 1975.* Berlin: Springer, 2013.

Davis, John. *The Great Exhibition.* Stroud: Sutton Publishing Ltd, 1999.

Davis, Lee N. *Corporate Alchemists: Power and Problems of the Chemical Industry.* Middlesex: Maurice Temple Smith Ltd, 1984.

De La Pena, Carolyn. *Empty Pleasures: The Study of Artificial Sweeteners from Sweet to Splenda.* Chapel Hill: University of North Carolina Press, 2010.

Debré, Patrice. *Louis Pasteur.* Translated by Elborg Forster. Baltimore: Johns Hopkins University Press, 2000.

Debus, Allen G. "Sir Thomas Browne and the Study of Colour Indicators." *Ambix* 10, 1 (February 1, 1962): 29–36.

Debus, Allen G. "Solution Analyses Prior to Robert Boyle." *Chymia* 8 (January 1962): 41–61.

Delanda, Manuel. *A New Philosophy of Society: Assemblage Theory and Social Complexity*. London: Continuum, 2006.

Delanghe, Joris R., and Marijn M. Speeckaert. "Creatinine Determination According to Jaffe—What Does It Stand For?" *NDT Plus* 4, 2 (2011): 83–86.

Deleuze, Gilles, and Felix Guattari. *A Thousand Plateaus*. London: Bloomsbury Academic, 2013.

Demortain, David. *Scientists and the Regulation of Risk: Standardising Control*. Cheltenham: Edward Elgar Publishing, 2011.

Desrosières, Alain. *The Politics of Large Numbers: A History of Statistical Reasoning*. Cambridge, MA: Harvard University Press, 2002.

Dessaux, Pierre-Antoine. "Chemical Expertise and Food Market Regulation in Belle-Epoque France." *History and Technology* 23, 4 (2007): 351–68.

Dingwall, Robert. *The Sociology of the Professions: Lawyers, Doctors, and Others*. New Orleans: Quid Pro Books, 2014.

Douglas, Mary. *Food in the Social Order: Mary Douglas: Collected Works*. Abingdon: Routledge, 2002.

Douglas, Mary. *Purity and Danger: An Analysis of Concepts of Pollution and Taboo*. Abingdon: Routledge, 2002.

Drouard, Alain, and Derek Oddy. *The Food Industries of Europe in the Nineteenth and Twentieth Centuries*. Abingdon: Routledge, 2016.

Drummond, J. C., and Anne Wilbraham. *The Englishman's Food: A History of Five Centuries of English Diet*. London: Pimlico, 1991.

Ellerbrock, Karl-Peter. "Lebensmittelqualität vor dem Ersten Weltkrieg: Industrielle Produktion und Staatliche Gesundheitspolitik." *Durchbruch zum Modernen Massenkonsum* (1987): 127–88.

Engel, Alexander. "Colouring Markets: The Industrial Transformation of the Dyestuffs Business Revisited." *Business History* 54, 1 (2012): 10–29.

Engel, Alexander. "Colouring the World: Marketing German Dyestuffs in the Late Nineteenth Century and Early Twentieth Centuries." In *Bright Modernity: Colour, Commerce, and Consumer Culture*, edited by Regina Lee Blaszczyk and Uwe Spiekermann, 37–55. Cham, Switzerland: Palgrave Macmillan, 2017.

Engel, Alexander, *Farben der Globalisierung: Die Enstehung moderner Märkte für Farbstoffe 1500–1900*. Frankfurt: Campus, 2009.

Ereshefsky, Marc. *The Poverty of the Linnaean Hierarchy: A Philosophical Study of Biological Taxonomy*. Cambridge: Cambridge University Press, 2000.

Eyler, J. M. "The Epidemiology of Milk-Borne Scarlet Fever: The Case of Edwardian Brighton." *American Journal of Public Health* 76, 5 (1986): 573–84.

Eyler, J. M. *Sir Arthur Newsholme and State Medicine, 1885–1935*. Cambridge: Cambridge University Press, 2002.

Fagin, Dan. *Tom's River: A Story of Science and Salvation*. New York: Bantam Dell Publishing Group, 2014.

Fairlie, Susan. *Dyestuffs in the Eighteenth Century*. Chichester: Economic History Society, 1965.

Feingold, Ben. "Hyperkinesis and Learning Disabilities Linked to Artificial

Food Flavors and Colors." *American Journal of Nursing* 75, 5 (1975): 797–800.

Feingold, Ben. *Why Your Child Is Hyperactive*. New York: Random House USA, Inc., 1988.

Fell, Ulricke, and Alan Rocke. "The Chemical Society of France in Its Formative Years, 1857–1914." In *Creating Networks in Chemistry: The Founding and Early History of Chemical Societies in Europe*, 91–112. Cambridge: Royal Society of Chemistry, 2008.

Ferrari, Matteo. *Risk Perception, Culture, and Legal Change*. Farnham, UK: Ashgate, 2009.

Ferrières, Madeleine. *Sacred Cow, Mad Cow*. New York: Columbia University Press, 2006.

Findling, John E., and Kimberly D. Pelle, eds. *Encyclopedia of World's Fairs and Expositions*. Jefferson, NC: McFarland, 2008.

Finn, Bernard S. *The History of Electrical Technology: An Annotated Bibliography*. New York: Garland Publishing, 1991.

Fischler, Claude. *L'homnivore: Le Goût, la Cuisine et le Corps*. Paris: O. Jacob, 1993.

Fitzgerald, Randall. *The Hundred-Year Lie: How to Protect Yourself from the Chemicals That Are Destroying Your Health*. New York: Penguin Group, 2007.

Fitzgerald, Robert. *Rowntree and the Marketing Revolution, 1862–1969*. Cambridge: Cambridge University Press, 1995.

Flandrin, Jean-Louis, and Massimo Montanari, eds. *Food: A Culinary History from Antiquity to the Present*. Translated by Albert Sonnenfeld. New York: Columbia University Press, 1999.

Fox, M. R. *Dye-Makers of Great Britain 1856–1976: A History of Chemists, Companies, Products, and Changes*. Liverpool: ICI Organics Division, 1987.

Fox, Robert, and Anna Guagnini, eds. *Education, Technology, and Industrial Performance in Europe, 1850–1939*. Cambridge: Cambridge University Press, 1993.

Fox, Robert, and Agustí Nieto-Galan. *Natural Dyestuffs and Industrial Culture in Europe, 1750–1880*. Cambridge: Science History Publications, 1999.

Fraser, W. Hamish. *The Coming of the Mass Market, 1850–1914*. London: Macmillan, 1981.

French, Michael, and Jim Phillips. *Cheated Not Poisoned?: Food Regulation in the United Kingdom, 1875–1938*. Manchester: Manchester University Press, 2000.

French, Michael, and Jim Phillips. "Sophisticates or Dupes? Attitudes toward Food Consumers in Edwardian Britain." *Enterprise and Society* 4, 3 (2003): 442–70.

French, Richard D. *Antivivisection and Medical Science in Victorian Society*. Princeton: Princeton University Press, 1975.

Fressoz, Jean-Baptiste. "Beck Back in the 19th Century: Towards a Genealogy of Risk Society." *History and Technology* 23, 4 (2007): 333–50.

Fressoz, Jean-Baptiste. "Gaz, Gazomètres, Expertises et Controverses: Londres, Paris, 1815–1860." *Le Courrier de l'environnement de l'INRA* 62 (2012): 31–56.

Fressoz, Jean-Baptiste, and Thomas Le Roux. "Protecting Industry and Commodifying the Environment: The Great Transformation of French Pollution Regulation, 1700–1840." In *Common Ground: Integrating the Social and Environmental in History*, edited by G. Massard-Guilbaud and S. Mosley, 340–66. Newcastle upon Tyne: Cambridge Scholars, 2011.

Fricke, John, Jacy Scarfone, and William Stillmann. *The Wizard of Oz: The Official 50th Anniversary Pictorial History*. New York: Grand Central Publishing, 1989.

Furlough, Ellen. *Consumer Cooperation in France: The Politics of Consumption, 1834–1930*. Ithaca, NY: Cornell University Press, 1991.

Furlough, Ellen, and Carl Strikwerda, eds. *Consumers Against Capitalism?* Lanham, MD: Rowman & Littlefield Publishers, 1999.

Fyfe, Aileen, and Bernard V. Lightman. *Science in the Marketplace: Nineteenth-Century Sites and Experiences*. Chicago: University of Chicago Press, 2007.

Fyke, Richard E. *The Bottle Book*. Salt Lake City: Peregrine Smith Books, 1987.

Gade, D. W. "Tradition, Territory, and Terroir in French Viniculture: Cassis, France, and Appellation Contrôlée." *Annals of the Association of American Geographers* 94 (2004): 848–67.

Gage, John. *Colour and Culture: Practice and Meaning from Antiquity to Abstraction*. London: Thames and Hudson, 1993.

Garfield, Simon. *Mauve: How One Man Invented a Colour That Changed the World*. London: Faber, 2001.

Gaskill, Nicholas. "Learning to see with Milton Bradley." In *Bright Modernity: Colour, Commerce, and Consumer Culture*, edited by Regina Lee Blaszczyk and Uwe Spiekermann, 55–77. Cham, Switzerland: Palgrave Macmillan, 2017.

Gaskill, Nicholas. "Vibrant Environments: The Feel of Color from the White Whale to the Red Wheelbarrow." Ph.D. diss., University of North Carolina, 2010.

Gee, Brian. "Amusement Chests and Portable Laboratories: Practical Alternatives to the Regular Laboratory." In *The Development of the Laboratory: Essays on the Place of Experiments in Industrial Civilization*, edited by Frank A. J. L. James, 37–59. Basingstoke, Hampshire: Palgrave Macmillan, 1989.

Gerber, Samuel M. *Chemistry and Crime: From Sherlock Holmes to Today's Courtroom*. Washington, DC: American Chemical Society, 1983.

Gernsheim, Alison. *Victorian and Edwardian Fashion: A Photographic Survey*. Mineola, NY: Courier Dover Publications, 1982.

Giddens, Anthony. *The Consequences of Modernity*. Cambridge, UK: Polity Press, 1991.

Giddens, Anthony. *Runaway World: How Globalization Is Reshaping Our Lives*. London: Profile Books, 2002.

Golan, Tal. *Laws of Men and Laws of Nature: The History of Scientific Expert*

Testimony in England and America. Cambridge, MA: Harvard University Press, 2007.

Golinski, Jan. *Science as Public Culture: Chemistry and Enlightenment in Britain, 1760–1820*. Cambridge: Cambridge University Press, 1999.

Good, Hal. "Methods of Measuring Food Color." *Food Quality Magazine*, January/February 2003.

Gooday, Graeme. "Liars, Experts, and Authorities." *History of Science* 46 (2008): 431–56.

Gooday, Graeme. *The Morals of Measurement: Accuracy, Irony, and Trust in Late Victorian Electrical Practice*. Cambridge: Cambridge University Press, 2004.

Gooding, David, Trevor Pinch, and Simon Schaffer. *The Uses of Experiment: Studies in the Natural Sciences*. Cambridge: Cambridge University Press, 1989.

Goody, Jack. *Cooking, Cuisine, and Class: A Study in Comparative Sociology*. Cambridge: Cambridge University Press, 1982.

Gradmann, Christoph. "Experimental Life and Experimental Disease: The Role of Animal Experiments in Robert Koch's Medical Bacteriology." *Futura* 18, 2 (2003): 80–88.

Gratzer, Walter. *Terrors of the Table: The Curious History of Nutrition.* Oxford: Oxford University Press, 2005.

Greer, Germaine. "Britain Doesn't Need Beet Sugar." *Telegraph*, June 30, 2013, accessed online June 1, 2015. https://www.telegraph.co.uk/foodanddrink/10145478/Germaine-Greer-Britain-doesnt-need-beet-sugar.html/.

Griffith, R. Marie. "Apostles of Abstinence: Fasting and Masculinity during the Progressive Era." *American Quarterly* 52, 4 (2000): 599–638.

Grivetti, Louis E., and Howard-Yana Shapiro. *Chocolate: History, Culture, and Heritage*. New York: John Wiley & Sons, 2011.

Grüne, Jutta. *Anfänge staatlicher Lebensmittelüberwachung in Deutschland: Der "Vater der Lebensmittelchemie" Joseph König*. Edited by Hans Jürgen Teuteberg. Stuttgart: Franz Steiner Verlag, 1994.

Guillem-Llobat, Ximo. "The Sugar Industry, Political Authorities, and Scientific Institutions in the Regulation of Saccharin: Valencia (1888–1939)." *Annals of Science* 68, 3 (2011): 401–24.

Haber, L. F. *Chemical Industry, 1900–30: International Growth and Technological Change*. Oxford: Oxford University Press, 1971.

Haber, L. F. *The Chemical Industry During the Nineteenth Century: A Study of the Economic Aspect of Applied Chemistry in Europe and North America*. Oxford: Clarendon Press, 1958.

Haines, Richard W. *Technicolor Movies: The History of Dye Transfer Printing*. Jefferson, NC: McFarland Co. Inc., 2003.

Halliday, Stephen. *The Great Stink of London: Sir Joseph Bazalgette and the Cleansing of the Victorian Metropolis*. Stroud: History Press Ltd, 2001.

Hamilton, Susan. *Animal Welfare and Anti-Vivisection 1870–1910: Frances Power Cobbe*. London: Taylor & Francis, 2004.

Hamlin, Christopher. *Public Health and Social Justice in the Age of Chadwick: Britain, 1800–1854*. Cambridge: Cambridge University Press, 1998.

Hamlin, Christopher. *A Science of Impurity: Water Analysis in Nineteenth Century Britain.* Berkeley: University of California Press, 1990.

Hamlin, Christopher. "Scientific Method and Expert Witnessing: Victorian Perspectives on a Modern Problem." *Social Studies of Science* 16, 3 (1986): 485–513.

Hammond, P. W., and Harold Egan. *Weighed in the Balance.* London: HMSO, 1992.

Hanan, Mohamed, Fathy Abd El-Wahab, and Gehan Salah El-Deen Moram. "Toxic Effects of Some Synthetic Food Colorants and/or Flavor Additives on Male Rats." *Toxicology and Industrial Health* 29, 2 (2013): 224–32.

Handbook of U.S. Colorants: Foods, Drugs, Cosmetics, and Medical Devices. New York: Wiley-Interscience, 1991.

Hannaway, Owen. *The Chemists and the Word: The Didactic Origins of Chemistry.* Baltimore: Johns Hopkins University Press, 1975.

Hanssen, Maurice. *E for Additives.* London: HarperCollins, 1984.

Hardy, Anne. *The Epidemic Streets: Infectious Disease and the Rise of Preventive Medicine, 1856–1900.* Oxford: Clarendon Press, 1993.

Hardy, Anne, and M. Eileen Magnello. "Statistical Methods in Epidemiology: Karl Pearson, Ronald Ross, Major Greenwood, and Austin Bradford Hill, 1900–1945." *Sozial- und Präventivmedizin* 47, 2 (2002): 80–89.

Haynes, Williams. *American Chemical Industry.* New York: Van Nostrand, 1945.

Heer, Jean. *Nestle: 125 Years, 1866–1991.* Vevey, Switzerland: Nestle, 1992.

Heick, Welf Henry. *A Propensity to Protect: Butter, Margarine and the Rise of Urban Culture in Canada.* Waterloo, ON: Wilfrid Laurier University Press, 1991.

Henderson, W. O. *The State and the Industrial Revolution in Prussia 1740–1870.* Liverpool: Liverpool University Press, 1967.

Hennessey, Rachel. "Living in Color: The Potential Dangers of Artificial Dyes." *Forbes*, August 2012, 27.

Hepler-Smith, Evan. "'Just as the Structural Formula Does': Names, Diagrams, and the Structure of Organic Chemistry at the 1892 Geneva Nomenclature Congress." *Ambix* 62, 1 (2015): 1–28.

Hierholzer, Vera. *Nahrung nach Norm: Regulierung von Nahrungsmittelqualität Nahrungsmittelqualität '"in der Industrialisierung 1871–1914.* Göttingen: Vandenhoeck & Ruprecht GmbH KG, 2010.

Hierholzer, Vera. "Searching for the Best Standard: Different Strategies of Food Regulation during German Industrialization." *Food and History* 5, 2 (2007): 295–318.

Hildebrandt, M. "The Trial of the Expert." *New Criminal Law Review* 10 (2007): 78–101.

Hilton, Matthew. *Consumerism in Twentieth-Century Britain: The Search for a Historical Movement.* Cambridge: Cambridge University Press, 2003.

Hirsh, Richard F. *Technology and Transformation in the American Electric Utility Industry.* Cambridge: Cambridge University Press, 2002.

Hisano, Ai. "Eye Appeal Is Buy Appeal: Business Creates the Color of Foods, 1870–1970." Ph.D. diss., University of Delaware, 2016.

Hisano, Ai. "The Rise of Synthetic Colours in the American Food Industry, 1870–1940." *Business History Review* 90, 3 (Autumn 2016): 483–504.

Hisano, Ai. *Visualizing Taste: How Business Changed the Look of What You Eat.* Cambridge, MA: Harvard University Press, 2019.

Hochheiser, Sheldon. "The Evolution of U.S. Food Colour Standards, 1913–1919." *Agricultural History* 55, 4 (1981): 385–91.

Hochheiser, Sheldon. "Synthetic Food Colors in the United States: A History under Regulation." Ph.D. diss., University of Wisconsin, 1982.

Hoegg, JoAndrea, and Joseph W. Alba. "Taste Perception: More than Meets the Tongue." *Journal of Consumer Research* 33, 4 (March 2007): 490–98.

Holloway, S. W. F. *Royal Pharmaceutical Society of Great Britain: A Political and Social History.* London: Pharmaceutical Press, 1991.

Holmes, Frederic L. "Beyond the Boundaries." In *Lavoisier in European Context: Negotiating a New Language for Chemistry,* edited by Bernadette Bensaude-Vincent, 267–78. Ann Arbor: University of Michigan Press, 1995.

Holmes, Frederic L. "The Complementarity of Teaching and Research in Liebig's Laboratory." *Osiris* 5 (1989): 121–64.

Homburg, Ernst. "The Emergence of Research Laboratories in the Dyestuffs Industry, 1870–1900." *British Journal for the History of Science* 25, 1 (1992): 91–111.

Homburg, Ernst. "The Influence of Demand on the Emergence of the Dye Industry: The Roles of Chemists and Colourists." *Journal of the Society of Dyers and Colourists* 99, 11 (1983): 325–33.

Homburg, Ernst. "The Rise of Analytical Chemistry and Its Consequences for the Development of the German Chemical Profession (1780–1860)." *Ambix* 46, 1 (1999): 1–32.

Homburg, Ernst. "Two Factions, One Profession: The Chemical Profession in German Society 1780–1870." In *The Making of the Chemist: The Social History of Chemistry in Europe, 1789–1914,* edited by David M Knight and Helge Kragh, 39–77. Cambridge: Cambridge University Press, 1998.

Homburg, Ernst, Anthony S. Travis, and Harm G. Schröter, eds. *The Chemical Industry in Europe, 1850–1914: Industrial Growth, Pollution, and Professionalization.* Dordrecht: Kluwer Academic, 1998.

Hopwood, Nick. *Haeckel's Embryos: Images, Evolution, and Fraud.* Chicago: University of Chicago Press, 2015.

Horrocks, Sally. "Consuming Science: Science, Technology and Food in Britain, 1870–1939." Ph.D. diss., University of Manchester, 1993.

Houghton, Walter E. *The Victorian Frame of Mind, 1830–1870.* New Haven: Yale University Press, 1963.

Howe, Anthony. *Free Trade and Liberal England, 1846–1946.* Oxford: Oxford University Press, 1998.

Hughes, Thomas Parker. *Networks of Power: Electrification in Western Society, 1880–1930.* Baltimore: John Hopkins University Press, 1993.

Hugill, Antony. *Sugar and All That: History of Tate and Lyle.* London: Gentry Books, 1978.

Hugill, Peter J., and Veit Bachmann. "The Route to the Techno-Industrial World Economy and the Transfer of German Organic Chemistry to

America Before, During, and Immediately After World War I." *Comparative Technology Transfer and Society* 3, 2 (2005): 158–86.

Humble, Nicola. *Culinary Pleasures: Cook Books and the Transformation of British Food*. London: Faber, 2006.

Hunt, Bruce J. "The Ohm Is Where the Art Is: British Telegraph Engineers and the Development of Electrical Standards." *Osiris* 2, 9 (1994): 48–63.

Hunt, Peter. *International Companion Encyclopedia of Children's Literature*. Abingdon: Routledge, 2004.

Hutchings, John B. *Food Color and Appearance*. Gaithersburg: Aspen, 1999.

Hutter, Bridget M. *Anticipating Risks and Organising Risk Regulation*. Cambridge: Cambridge University Press, 2010.

Hutter, Bridget M. *Managing Food Safety and Hygiene: Governance and Regulation as Risk Management*. Cheltenham: Edward Elgar Publishing, 2011.

Huxley, Leonard, ed. *The Life and Letters of Thomas Henry Huxley*. Vol 1. London 1900.

Hyman, Gwen. *Making a Man: Gentlemanly Appetites in the Nineteenth-Century British Novel*. Athens: Ohio University Press, 2009.

Inkster, Ian, Graeme Gooday, and James Sumner, eds. *History of Technology*. Vol. 28. Special issue: *By Whose Standards? Standardization, Stability and Uniformity in the History of Information and Electrical Technologies*.

Inkster, Ian, and Jack Morrell. *Metropolis and Province: Science in British Culture, 1780–1850*. Philadelphia: University of Pennsylvania Press, 1983.

James, Frank A. J. L. *The Development of the Laboratory: Essays on the Place of Experiments in Industrial Civilization*. Basingstoke, Hampshire: Palgrave Macmillan, 1989.

Jardine, N., J. A. Secord, and E. C. Spary. *Cultures of Natural History*. Cambridge: Cambridge University Press, 1996.

Jasanoff, Sheila. "Breaking the Waves in Science Studies: Comment on H. M. Collins and Robert Evans, 'The Third Wave of Science Studies.'" *Social Studies of Science* 33, 3 (2003): 389–400.

Jefferys, James B. *Retail Trading in Britain, 1850–1950*. Cambridge: Cambridge University Press, 1954.

Johnson, Jeffrey A. "Academic Self-Regulation and the Chemical Profession in Imperial Germany." *Minerva* 23, 2 (1985): 241–71.

Johnson, Jeffrey A. "Germany: Discipline-Industry-Profession; German Chemical Organisations 1867–1914." In *Creating Networks in Chemistry: The Founding and Early History of Chemical Societies in Europe*, edited by Anita Kildebæk Nielsen and Soňa Štrbáňová, 113–38. Cambridge: Royal Society of Chemistry, 2008.

Johnston, Sean F. *A History of Light and Colour Measurement: Science in the Shadows*. Bristol: CRC Press, 2001.

Jones, Carol A. G. *Expert Witnesses: Science, Medicine, and the Practice of Law*. Oxford: Clarendon Press, 1994.

Jones, William Jervis. *German Colour Terms: A Study in Their Historical Evolution from Earliest Times to the Present*. Amsterdam: John Benjamins Publishing, 2013.

Kahl, Thomas. "Aniline." In *Ullmann's Encylcopedia of Industrial Chemistry*, 161–68. New York: John Wiley & Sons, 2007.

Kargon, Robert. "Expert Testimony in Historical Perspective." *Law and Human Behavior* 10, 1–2 (1986): 15–27.

Kassim, L. "The Co-operative Movement and Food Adulteration in the Nineteenth Century." *Manchester Region History Review* 15 (2001): 9–18.

Keane, Melanie *Science in Wonderland*. Oxford: Oxford University Press, 2015.

Kelley, Victoria. *Soap and Water: Cleanliness, Dirt, and the Working Classes in Victorian and Edwardian Britain*. London: I. B. Tauris & Co. Ltd., 2010.

Kennaway, Ernest. "The Identification of a Carcinogenic Compound in Coal-Tar." *British Medical Journal* 2, 4942 (1955): 749–52.

Keyser, Catherine. *Artificial Color: Modern Food and Racial Fictions*. Oxford: Oxford University Press, 2019.

Kirchelle, Claas. "Toxic Confusion: The dilemma of Antibiotic Regulation in West German Food Production (1951–1990)." *Endeavour* 2 (2016): 114–27.

Klein, Ursula. *Experiments, Models, Paper Tools: Cultures of Organic Chemistry in the Nineteenth Century*. Stanford: Stanford University Press, 2003.

Klein, Ursula. "Technoscience avant la Lettre." *Perspectives on Science* 13, 2 (2005): 226–66.

Klein, Ursula, and Wolfgang Lefèvre. *Materials in Eighteenth-Century Science: A Historical Ontology*. Cambridge, MA: MIT Press, 2007.

Klein, Ursula, and Carsten Reinhardt. *Objects of Chemical Inquiry*. Sagamore Beach, MA: Science History Publications, 2014.

Klein, Ursula, and E. C. Spary, eds. *Materials and Expertise in Early Modern Europe: Between Market and Laboratory*. Chicago: University of Chicago Press, 2010.

Knight, David M, and Helge Kragh. *The Making of the Chemist: The Social History of Chemistry in Europe, 1789–1914*. Cambridge: Cambridge University Press, 1998.

Koerner, Lisbet. *Linnaeus: Nature and Nation*. Cambridge, MA: Harvard University Press, 2009.

Kortsch, Christine Bayles. *Dress Culture in Late Victorian Women's Fiction: Literacy, Textiles, and Activism*. Farnham: Ashgate Publishing, Ltd., 2009.

Krätz, Otto. "Der Chemiker in den Gründerjahren." In *Der Chemiker im Wandel der Zeiten: Skizzen zur geschichtlichen Entwicklung des Berufsbildes*, edited by Eberhard Schmauderer, 259–85. Weinheim, Germany: Wiley-VCH, 1973.

Krislov, Samuel. *How Nations Choose Product Standards and Standards Change Nations*. Pittsburgh: University of Pittsburgh Press, 1997.

Kuehni, Rolf G. "The Early Development of the Munsell System." *Color Research and Application* 27, 1 (2002): 20–27.

Kumar, Prakesh. *Indigo Plantations and Science in Colonial India*. New York: Cambridge University Press, 2012.

Kumar, Prakesh. "Plantation Indigo and Synthetic Indigo: Redefinition of a

Colonial Commodity." *Comparative Studies in Society and History* 58, 2 (April 2016): 407–31.

Lacassagne, Antoine. "Kennaway and the Carcinogens." *Nature* 191, 4790 (1961): 743–47.

Landa, Edward, and Mark Fairchild. "Charting Color from the Eye of the Beholder." *American Scientist* 93, 5 (2005).

Landsman, Stephan. "Of Witches, Madmen, and Products Liability: An Historical Survey of the Use of Expert Testimony." *Behavioral Sciences and the Law* 13, 2 (1995): 131–57.

Latour, Bruno. *Reassembling the Social: An Introduction to Actor-Network Theory*. Oxford: Oxford University Press, 2007.

Latour, Bruno. *Science in Action: How to Follow Scientists and Engineers Through Society*. Cambridge, MA: Harvard University Press, 1987.

Latour, Bruno, and Steve Woolgar. *Laboratory Life: The Construction of Scientific Facts*. Edited by Jonas Salk. Princeton: Princeton University Press, 1986.

Layton, James, and David Pierce. *The Dawn of Technicolor, 1915–1935*. New York: Distributed Art Publishers, 2015.

Leapman, Michael. *The World for a Shilling: How the Great Exhibition of 1851 Shaped a Nation*. London: Headline Book Publishing, 2001.

Lenoir, Timothy. "Revolution from Above: The Role of the State in Creating the German Research System, 1810–1910." *American Economic Review* 88, 2 (1998): 22–27.

Lesch, John E. *Science and Medicine in France: The Emergence of Experimental Physiology, 1790–1855*. Cambridge, MA: Harvard University Press, 2013.

Leslie, Esther. *Synthetic Worlds: Nature, Art, and the Chemical Industry*. London: Reaktion Books, 2005.

Levenstein, Harvey. *Fear of Food: A History of Why We Worry About What We Eat*. Chicago: University of Chicago Press, 2012.

Levenstein, Harvey. *Revolution at the Table: The Transformation of the American Diet*. Berkeley: University of California Press, 2003.

Lévi-Strauss, Claude. *Introduction to a Science of Mythology*. London: Cape, 1978.

Lightman, Bernard. *Victorian Science in Context*. Chicago: University of Chicago Press, 1997.

Lyman, B. *A Psychology of Food: More than a Matter of Taste*. Berlin: Springer, 2012.

MacDonagh, Oliver. *Early Victorian Government, 1830–1870*. London: Weidenfeld and Nicolson, 1977.

MacLeod, Roy M. *Government and Expertise: Specialists, Administrators, and Professionals 1860–1919*. Cambridge: Cambridge University Press, 1988.

Marcus, Alan I. "Setting the Standard: Fertilizers, State Chemists, and Early National Commercial Regulation, 1880–1887." *Agricultural History* 61, 1 (1987): 47–73.

Mari, L. "Epistemology of Measurement." *Measurement* 34, 1 (2003): 17–30.

Marquardt, Hans, Siegfried G. Schäfer, Roger O. McClellan, and Frank Welsch. *Toxicology*. Cambridge, MA: Academic Press, 1999.

McCann, Donna, Angelina Barrett, Alison Cooper, Debbie Crumpler, Linda Dalen, Kate Grimshaw, et al. "Food Additives and Hyperactive Behaviour in 3-Year-Old and 8/9-Year-Old Children in the Community: A Randomised, Double-Blinded, Placebo-Controlled Trial." *Lancet*. 370, 9598 (November 3, 2007): 1560–67.

McClelland, Charles E. *The German Experience of Professionalization: Modern Learned Professions and Their Organizations from the Early Nineteenth Century to the Hitler Era*. Cambridge: Cambridge University Press, 2002.

Medlin, Sophie. "Activated Charcoal Doesn't Detox the Body." *theconversation.com*, June 12, 2018.

Meinel, Christoph, and Hartmut Scholz, eds. *Die Allianz von Wissenschaft und Industrie: August Wilhelm Hofmann (1818–1892): Zeit, Werk, Wirkung*. Weinheim: Wiley VCH, 1992.

Mennell, Stephen. *All Manners of Food: Eating and Taste in England and France from the Middle Ages to the Present*. Urbana: University of Illinois Press, 1996.

Mennell, Stephen, Anne Murcott, Anneke H. van Otterloo, and International Sociological Association. *The Sociology of Food: Eating, Diet, and Culture*. New York: Sage, 1992.

Merki, Christoph Maria. *Zucker gegen Saccharin: Zur Geschichte der künstlichen Süssstoffe*. Frankfurt: Campus, 1993.

Meyer-Renschhausen, Elisabeth, and Albert Wirz. "Dietetics, Health Reform, and Social Order: Vegetarianism as a Moral Physiology; The Example of Maximilian Bircher-Benner (1867–1939)." *Medical History* 43, 3 (1999): 323–41.

Miller, Geoffrey P. "Public Choice at the Dawn of the Special Interest State: The Story of Butter and Margarine." *California Law Review* 77, 1 (1989): 83–131.

Miller, Ian. *A Modern History of the Stomach: Gastric Illness, Medicine, and British Society, 1800–1950*. Abingdon: Routledge, 2015.

Miller, Melanie. "Food Colours: A Study of the Effects of Regulation." Ph.D. diss., Aston University, 1987.

Minard, Gayle. "The History of Surgically Placed Feeding Tubes." *Nutrition in Clinical Practice* 21, 6 (2006): 626–33.

Mintz, Sidney Wilfred. "The Changing Roles of Food in the Study of Consumption." In *Consumption and the World of Goods*, edited by John Brewer and Roy Porter, 261–73. New York: Routledge, 1993.

Mintz, Sidney Wilfred. *Sweetness and Power: The Place of Sugar in Modern History*. New York: Viking Penguin, 1985.

Mintz, Sidney Wilfred. *Tasting Food, Tasting Freedom: Excursions into Eating, Culture, and the Past*. Boston: Beacon Press, 1996.

Money, John. *The Destroying Angel: Sex, Fitness, and Food in the Legacy of Degeneracy Theory: Graham Crackers, Kellogg's Corn Flakes, and American Health History*. New York: Prometheus Books, 1985.

Morabia, Alfredo, ed. *A History of Epidemiologic Methods and Concepts.* Basel: Birkhäuser Basel, 2006.

Morgan, Miriam. "Sugar, Sugar: Cane and Beet Share the Same Chemistry but Act Differently in the Kitchen." *SFGATE (San Francisco Chronicle* website), March 31, 1999. Accessed June 1, 2015.

Morris, Peter J. T. *The Matter Factory: A History of the Chemistry Laboratory.* London: Reaktion Books, 2015.

Morris, Peter J. T., and Anthony S. Travis. "The Chemical Society of London and the Dye Industry in the 1860s." *Ambix* 39, 3 (1992): 117–26.

Murmann, Johann Peter. "Knowledge and Competitive Advantage in the Synthetic Dye Industry, 1850–1914: The Coevolution of Firms, Technology, and National Institutions in Great Britain, Germany, and the United States." *Enterprise and Society* 1, 4 (2000): 699–704.

Murmann, Johann Peter. *Knowledge and Competitive Advantage: The Co-evolution of Firms, Technology, and National Institutions.* Cambridge: Cambridge University Press, 2003.

Neswald, Elizabeth. "Francis Gano Benedict's Reports of Visits to Foreign Laboratories and the Carnegie Nutrition Laboratory." *Actes d'Història de la Ciència I de la Tècnica* 4 (2012): 11–32.

Nickerson, Dorothy. "History of the Munsell Color System, Company, and Foundation." *Color Research and Application* 1, 3 (1976): 121–30.

Nicklas, Charlotte. "New Words and Fanciful Names: Dyes, Color, and Fashion in the Mid-Nineteenth Century." In *Bright Modernity: Color, Commerce, and Consumer Culture,* edited by Regina Lee Blaszczyk and Uwe Spiekermann, 97–111. Cham, Switzerland: Palgrave Macmillan, 2017.

Nicklas, Charlotte. "Splendid Hues: Colour, Dyes, Everyday Science, and Women's Fashion, 1840–1875." Ph.D. diss., University of Brighton, 2009.

Nieto-Galan, Agustí. *Colouring Textiles: A History of Natural Dyestuffs in Industrial Europe.* Dordrecht: Kluwer Academic, 2001.

Nissenbaum, Stephen. *Sex, Diet, and Debility in Jacksonian America: Sylvester Graham and Health Reform.* Westport, CT: Greenwood Press, 1980.

Noakes, Richard. "*Punch* and Comic Journalism in Mid-Victorian Britain." In Geoffrey Cantor, Gowan Dawson, Graeme Gooday, Richard Noakes, Sally Shuttleworth, and Jonathan R. Topham, *Science in the Nineteenth-Century Periodical: Reading the Magazine of Nature,* 91–122. Cambridge: Cambridge University Press, 2004.

Oddy, Derek J. "Food Quality in London and the Rise of the Public Analyst, 1870–1939." In *Food and the City in Europe since 1800,* edited by Peter Atkins, Peter Lummel, and Derek J. Oddy. Farnham: Ashgate, 2007.

Okun, Mitchell. "Fair Play in the Marketplace: Adulteration and the Origins of Consumerism." Ph.D. diss., City University of New York, 1983.

Okun, Mitchell. *Fair Play in the Marketplace: The First Battle for Pure Food and Drugs.* DeKalb: Northern Illinois University Press, 1986.

Olson, Richard. *Science and Scientism in Nineteenth-Century Europe.* Urbana: University of Illinois Press, 2008.

Orland, Barbara. "The Chemistry of Everyday Life." In *Communicating Chemistry: Textbooks and Their Audiences, 1789–1939,* edited by Anders

Lundgren and Bernadette-Bensaude Vincent, 327–67. Canton, MA: Science History Publications, 2000.

O'Rourke, Kevin. "British Trade Policy in the 19th Century: A Review Article." Working Paper. College Dublin, Department of Political Economy, 1999.

Otterloo, A. H. van. "The Development of Public Distrust of Modern Food Technology in the Netherlands." In *Food Technology, Science, and Marketing*, edited by A. P. den Hartog, 253–67. East Linton: Tuckwell Press, 1995.

Paradis, James G. *Victorian Science and Victorian Values: Literary Perspectives.* New Brunswick, NJ: Rutgers University Press, 1981.

Partington, James Riddick. *A History of Chemistry.* London: Macmillan, 1964.

Patel, Raj. *Stuffed and Starved.* London: Portobello Books, 2007.

Paul, Harry W. *From Knowledge to Power: The Rise of the Science Empire in France, 1860–1939.* Cambridge: Cambridge University Press, 2003.

Paulus, Ingeborg. *Search for Pure Food: A Sociology of Legislation in Britain.* London: Martin Robinson, 1974.

Petrick, Gabriella M. "The Industrialization of Food." In *The Oxford Handbook of Food History*, ed. Jeffrey M. Pilcher, 258–78. Oxford: Oxford University Press, 2012.

Petrick, Gabriella M. "'Purity as Life': H. J. Heinz, Religious Sentiment, and the Beginning of the Industrial Diet." *History and Technology* 27, 1 (2011): 37–64.

Phillips, Denise. *Acolytes of Nature: Defining Natural Science in Germany, 1770–1850.* Chicago: University of Chicago Press, 2012.

Phillips, Jim, and Michael French. "Adulteration and Food Law, 1899–1939." *Twentieth Century British History* 9, 3 (1998): 350–69.

Pickering, Andrew. "Decentering Sociology: Synthetic Dyes and Social Theory." *Perspectives on Science* 13, 3 (2005): 352–405.

Pickering, Andrew. *The Mangle of Practice: Time, Agency, and Science.* Chicago: University of Chicago Press, 1995.

Pickering, Andrew. *Science as Practice and Culture.* Chicago: University of Chicago Press, 1992.

Pickstone, John V. *Ways of Knowing: A New History of Science, Technology, and Medicine.* Manchester: Manchester University Press, 2000.

Pinch, Trevor. "Towards an Analysis of Scientific Observation: The Externality and Evidential Significance of Observational Reports in Physics." *Social Studies of Science* 15, 1 (1985): 3–36.

Pollan, Michael. *Cooked: A Natural History of Transformation.* London: Penguin Books, 2013.

Pollan, Michael. *The Omnivore's Dilemma.* London: Penguin Books, 2006.

Porter, Dorothy. *Health, Civilization, and the State: A History of Public Health from Ancient to Modern Times.* London: Routledge, 1999.

Porter, Roy. *The Greatest Benefit to Mankind: A Medical History of Humanity from Antiquity to the Present.* London: Fontana Press, 1999.

Porter, Theodore M. *Karl Pearson: The Scientific Life in a Statistical Age.* Princeton: Princeton University Press, 2010.

Porter, Theodore M. *Trust in Numbers: The Pursuit of Objectivity in Science and Public Life.* Princeton: Princeton University Press, 1996.

Powers, Michael Brian. "The Early Industrial Achievements of the Schoelkopf Family." MA thesis, Niagara University, 1979.

Price, Paul J., William A. Suk, Aaron E. Freeman, William T. Lane, Robert L. Peters, Mina Lee Vernon, and Robert J. Huebner. "In Vitro and in Vivo Indications of the Carcinogenicity and Toxicity of Food Dyes." *International Journal of Cancer* 21, 3 (1978): 361–67.

Principe, Laurence M. *The Secrets of Alchemy*. Chicago: University of Chicago Press, 2013.

Proctor, Robert N. *The Nazi War on Cancer*. New ed. Princeton: Princeton University Press, 2000.

Rabier, Christelle. *Fields of Expertise: A Comparative History of Expert Procedures in Paris and London, 1600 to Present*. Newcastle, UK: Cambridge Scholars Publishing, 2007.

Reinhardt, Carsten. *Heinrich Caro and the Creation of Modern Chemical Industry*. Berlin: Springer, 2000.

Reynolds, Terry S. "Defining Professional Boundaries: Chemical Engineering in the Early 20th Century." *Technology and Culture* 27, 4 (1986): 694–716.

Rioux, Sébastien. "Capitalist Food Production and the Rise of Legal Adulteration: Regulating Food Standards in 19th Century Britain." *Journal of Agrarian Change* 19, 1 (2019): 64–81.

Roberts, Gerrylynn K. "Bridging the Gap between Science and Practice: The English Years of August Wilhelm Hofmann, 1845–1865." In *Die Allianz von Wissenschaft und Industrie: August Wilhelm Hofmann (1818–1892)*, edited by Christoph Meinel and Hartmut Scholz, 89–99. Weinheim: VCH Publishers, 1992.

Roberts, Lissa. "The Death of the Sensuous Chemist: The 'New' Chemistry and the Transformation of Sensuous Technology." *Studies in History and Philosophy of Science Part A* 26, 4 (1995): 503–29.

Roberts, Lissa. "Filling the Space of Possibilities: Eighteenth-Century Chemistry's Transition from Art to Science." *Science in Context* 6, 2 (1993).

Rocke, Alan J. *Image and Reality: Kekulé, Kopp, and the Scientific Imagination*. Chicago: University of Chicago Press, 2010.

Rocke, Alan J. *Nationalizing Science: Adolphe Wurtz and the Battle for French Chemistry*. Cambridge: MIT Press, 2000.

Rocke, Alan J. *The Quiet Revolution: Hermann Kolbe and the Science of Organic Chemistry*. Berkeley: University of California Press, 1993.

Rosenfeld, Louis. "Gastric Tubes, Meals, Acid, and Analysis: Rise and Decline." *Clinical Chemistry* 43, 5 (1997): 837–42.

Rossi, Michael. "Let's Go Color Shopping with Charles Sanders Peirce: Color Scientists as Consumers of Color." In *Bright Modernity: Color, Commerce, and Consumer Culture*, edited by Regina Lee Blaszczyk and Uwe Spiekermann, 113–33. Cham, Switzerland: Palgrave Macmillan, 2017.

Rowan, Andrew N. *Of Mice, Models, and Men: A Critical Evaluation of Animal Research*. Albany: SUNY Press, 1984.

Royal Society of Chemistry. *The Fight Against Food Adulteration*. London: Royal Society of Chemistry, 2005.

Rubery, Matthew. *The Novelty of Newspapers: Victorian Fiction After the Invention of the News*. Oxford: Oxford University Press, 2009.

Rudwick, Martin J. S. *Earth's Deep History: How It Was Discovered and Why It Matters*. Chicago: University of Chicago Press, 2014.

Rüschemeyer, Dietrich. "Professional Autonomy and the Social Control of Expertise." *The Sociology of the Professions: Lawyers, Doctors, and Others*, edited by Robert Dingwall and Philip Lewis, 38–58. New Orleans: Quid Pro Books, 2014.

Ruske, Walter. *100 Jahre Deutsche Chemische Gesellschaft*. Weinheim: Deutsche Chemische Gesellschaft, 1967.

Russell, Colin A. *Chemists by Profession: The Origins and Rise of the Royal Institute of Chemistry*. Milton Keynes: Open University Press for the Institute, 1977.

Russell, Colin A. *Edward Frankland: Chemistry, Controversy, and Conspiracy in Victorian England*. Cambridge: Cambridge University Press, 2003.

Samuelson, J. *The Civilisation of Our Day: Essays*. London: S. Low, Marston and Co., 1896.

Schaeffer, Albert. *Die Entwicklung der künstlichen organischen Farbstoffe*. Hofheim am Taunus, Germany: A. Schaeffer, 1951.

Schaeffer, Juliann. "Color Me Healthy: Eating for a Rainbow of Benefits." *Today's Dietician* 10, 11 (2008): 34.

Schaffer, Simon. "Astronomers Mark Time: Discipline and the Personal Equation." *Science in Context* 2 (1988): 115–45.

Schaffer, Simon. "Late Victorian Metrology and Its Instrumentation: A Manufactory of Ohms." In *Invisible Connections: Instruments, Institutions, and Science*, edited by Robert Bud, 23–56. Bellingham: SPIE, 1992.

Schaffer, Simon. "Metrology, Metrication, and Victorian Values." In *Victorian Science in Context*, edited by Bernard Lightman, 438–76. Chicago: University of Chicago Press, 1997.

Shteir, Ann B. *Cultivating Women, Cultivating Science: Flora's Daughters and Botany in England, 1760 to 1860*. Baltimore: Johns Hopkins University Press, 1999.

Schmauderer, Eberhard. *Der Chemiker im Wandel der Zeiten: Skizzen zur geschichtlichen Entwicklung des Berufsbildes*. Weinheim/Bergstr.: Wiley-VCH, 1973.

Schmidt, Susanne K. *Coordinating Technology: Studies in the International Standardization of Telecommunications*. Cambridge, MA: MIT Press, 1998.

Scholliers, Peter. "Defining Food Risks and Food Anxieties Throughout History." *Appetite* 51, 1 (2008): 3–6.

Scholliers, Peter, and Patricia Van den Eeckhout. "Hearing the Consumer? The Laboratory, the Public, and the Construction of Food Safety in Brussels (1840s–1910s)." *Journal of Social History* 44, 4 (2011): 1139–55.

Schoonhoven, Claudia Bird, and Elaine Romanelli. *The Entrepreneurship Dynamic: Origins of Entrepreneurship and the Evolution of Industries*. Stanford: Stanford University Press, 2001.

Schröter, Harm G., and Anthony S. Travis. "An Issue of Different Mentalities: National Approaches to the Development of the Chemical Industry in Britain and Germany Before 1914." In *The Chemical Industry in Europe, 1850–1914: Industrial Growth, Pollution, and Professionalization*, edited by Ernst Homburg, Anthony S. Travis, and Harm G. Schröter, 95–120. Dordrecht: Kluwer Academic, 1998.

Schug, Thaddeus, Amanda Janesick, Bruce Blumberg, and Jerrold J. Heindel. "Endocrine Disrupting Chemicals and Disease Susceptibility." *Journal of Steroid Biochemistry and Molecular Biology* 127, 2–5 (2011): 204–5.

Schwartz, Hillel. *Never Satisfied: A Cultural History of Diets, Fantasies, and Fat*. New York: Free Press; London: Collier Macmillan, 1986.

Secord, James. "Knowledge in Transit." *Isis* 95, 4 (2004): 654–72.

Setbon, Michel, Jocelyn Raude, Claude Fischler, and Antoine Flahault. "Risk Perception of the 'Mad Cow Disease' in France: Determinants and Consequences." *Risk Analysis* 25, 4 (2005): 813–26.

Shapin, Steven. "Science and the Public." In *Companion to the History of Modern Science*, edited by G. N. Cantor, 990–1007. Abingdon: Routledge, 1996.

Shapin, Steven. *A Social History of Truth: Civility and Science in Seventeenth-Century England*. Chicago: University of Chicago Press, 1995.

Shapin, Steven, and Simon Schaffer. *Leviathan and the Air-Pump: Hobbes, Boyle, and the Experimental Life*. Princeton: Princeton University Press, 2011 [1989].

Shaw, Gareth. "Changes in Consumer Demand and Food Supply in Nineteenth-Century British Cities." *Journal of Historical Geography* 11, 3 (1985): 280–96.

Sheets-Pyenson, Susan. "Popular Science Periodicals in Paris and London: The Emergence of a Low Scientific Culture, 1820–1875." *Annals of Science* (November 1985): 549–72.

Silver, G. A. "Virchow, the Heroic Model in Medicine: Health Policy by Accolade." *American Journal of Public Health* 77, 1 (1987): 82–88.

Simon, Christian. "The Swiss Chemical Industry." In *The Chemical Industry in Europe, 1850–1914: Industrial Growth, Pollution, and Professionalization*, edited by Ernst Homburg, Anthony S. Travis, and Harm G. Schröter, 9–27. Dordrecht: Kluwer Academic, 1998.

Simon, Jonathan. *Chemistry, Pharmacy, and Revolution in France, 1777–1809*. Farnham: Ashgate Publishing, Ltd., 2013.

Simons Slater, Katharine. "Little Geographies: Children's Literature and Local Place." Ph.D. diss., University of California, 2013.

Smith, Andrew F. *Pure Ketchup: A History of America's National Condiment, with Recipes*. Columbia: University of South Carolina Press, 1996.

Smith, Michael Stephen. *The Emergence of Modern Business Enterprise in France, 1800–1930*. Cambridge, MA: Harvard University Press, 2006.

Spary, E. C. *Eating the Enlightenment: Food and the Sciences in Paris*. Chicago: University of Chicago Press, 2012.

Spary, E. C. *Feeding France: New Sciences of Food, 1760–1815*. Cambridge: Cambridge University Press, 2014.

Spary, E. C. "Ways with Food." *Journal of Contemporary History* 40, 4 (2005): 763–71.

Spence, Charles. "On the Psychological Impact of Food Colour." *Flavour* 4, 1 (2015): 1–16.

Spencer, Dianne. "Choose Your Food by Colour—and Lose Weight!" *Daily Mail Online*, accessed October 5, 2016. http://www.dailymail.co.uk/health/article-53080/Choose-food-colour--lose-weight.html.

Spiekermann, Uwe. "Redefining Food: The Standardization of Products and Production in Europe and the United States, 1880–1914." *History and Technology* 27, 1 (2011): 11–36.

Spiekermann, Uwe. "Twentieth-Century Product Innovations in the German Food Industry." *Business History Review* 83, 2 (2009): 291–315.

Stanziani, Alessandro. "Information, Quality, and Legal Rules: Wine Adulteration in Nineteenth Century France." *Business History* 51, 2 (2009): 268–91.

Stanziani, Alessandro. "La Mesure de la Qualité du Vin en France, 1871–1914." *Food and History* 2, 1 (2004): 191–226.

Stanziani, Alessandro. "Municipal Laboratories and the Analysis of Foodstuffs in France under the Third Republic." In *Food and the City in Europe since 1800*, edited by Peter Atkins, Peter Lummel, and Derek J. Oddy. Farnham: Ashgate, 2007.

Stanziani, Alessandro. "Negotiating Innovation in a Market Economy: Foodstuffs and Beverages Adulteration in Nineteenth-Century France." *Enterprise and Society* 8, 2 (2007): 375–412.

Stanziani, Alessandro. *Rules of Exchange: French Capitalism in Comparative Perspective, Eighteenth to Early Twentieth Centuries.* Cambridge: Cambridge University Press, 2012.

Star, Susan Leigh, and James R. Griesemer. "Institutional Ecology, 'Translations,' and Boundary Objects: Amateurs and Professionals in Berkeley's Museum of Vertebrate Zoology, 1907–39." *Social Studies of Science* 19, 3 (1989): 387–420.

Stare, F. J., E. M. Whelan, and M. Sheridan. "Diet and Hyperactivity: Is There a Relationship?" *Pediatrics* 66, 4 (October 1980): 521–25.

Steen, Juliette. "So, Eating 'Activated Charcoal' Is a Thing Now: Just Another Wacky Trend?" *Huffpost*, March 11, 2016. https://www.huffingtonpost.com.au/2016/11/02/so-eating-activated-charcoal-is-a-thing-now_a_21596704/.

Steere-Williams, Jacob. "A Conflict of Analysis: Analytical Chemistry and Milk Adulteration in Victorian Britain." *Ambix* 61, 3 (2014): 279–98.

Steere-Williams, Jacob. "Lacteal Crises: Debates over Milk Purity in Victorian England." In *Victorian Medicine and Popular Culture*, edited by Louise Penner and Tabitha Sparks, 53–66. London: Pickering & Chatto, 2015.

Steere-Williams, Jacob. "The Perfect Food and the Filth Disease: Milk-Borne Typhoid and Epidemiological Practice in Late Victorian Britain." *Journal of the History of Medicine and Allied Sciences* 65, 4 (2010): 514–45.

Stern, Rebecca F. "'Adulterations Detected': Food and Fraud in Christina

Rossetti's 'Goblin Market.'" *Nineteenth-Century Literature* 57, 4 (2003): 477–511.

Stern, Rebecca F. *Home Economics: Domestic Fraud in Victorian England.* Columbus: Ohio State University Press, 2008.

Steward, H. F. "Advisory and Adversary Processes in the Assessment and Control of Food Additives." Ph.D. diss., University of Manchester, 1978.

Stoff, Heiko, *Gift in der Nahrung: Zur Genese der Verbraucherpolitik Mitte des 20 Jahrhunderts* Stuttgart: Frans Steiner Verlag, 2015.

Stuyvenberg, J. H. Van, ed. *Margarine: An Economic, Social and Scientific History, 1869–1969.* Liverpool: Liverpool University Press, 1969.

Sumner, James. "Retailing Scandal: The Disappearance of Friedrich Accum." In *(Re)creating Science in Nineteenth-Century Britain: An Interdisciplinary Approach*, edited by Amanda Mordavsky Caleb, 32–48. Newcastle, UK: Cambridge Scholars Publishing, 2007.

Szabadváry, Ferenc. *History of Analytical Chemistry.* Yverdon, Switzerland: Gordon and Breach Science Publishers, 1992.

Tal, Eran. "Old and New Problems in Philosophy of Measurement." *Philosophy Compass* 8, 12 (2013): 1159–73.

Taussig, Michael T. *What Color Is the Sacred?* Chicago: University of Chicago Press, 2009.

Taylor, F. Sherwood. *A History of Industrial Chemistry.* London: Heinemann, 1957.

Tenhoor, Meredith. "Architecture and Biopolitics at Les Halles." *French Politics, Culture, and Society* 25, 2 (Summer 2007): 73–92.

Teughels, Nelleke, and Peter Scholliers. *A Taste of Progress: Food at International and World Exhibitions in the Nineteenth and Twentieth Centuries.* London: Routledge, 2016.

Teuteberg, Hans Jürgen. "Adulteration of Food and Luxuries and the Origins of Uniform State Food Legislation in Germany." *Zeitschrift für Ernährungswissenschaft* 34, 2 (1995): 95–112.

Teuteberg, Hans Jürgen. *European Food History: A Research Review.* Leicester: Leicester University Press, 1992.

Timmermans, Stefan, and Steven Epstein. "A World of Standards but Not a Standard World: Toward a Sociology of Standards and Standardization." *Annual Review of Sociology* 36, 1 (2010): 69–89.

Todes, Daniel P. *Pavlov's Physiology Factory: Experiment, Interpretation, Laboratory Enterprise.* Baltimore: John Hopkins University Press, 2001.

Tomic, Sacha. *Aux origines de la chimie organique.* Rennes: Université de Rennes, 2010.

Tomic, Sacha. "The Toxicological Laboratory of Paris During Jules Ogier's Direction 1883–1911." Presented at the 8th European Spring School on History of Science and Popularization | Societat Catalana d'Història de la Ciència ide la Tècnica, July 1, 2015, Menorca.

Tomic, Sacha, and X. Guillem-Llobat. "New Sites for Food Quality Surveillance in European Centres and Peripheries: To What Extent Was the Municipal Laboratory of Paris a Model for Iberian Laboratories?" Presented at the Sites of Chemistry in the 19th Century, Valencia, July 6, 2012.

Travis, Anthony S. "Between Broken Root and Artificial Alizarin: Textile Arts and Manufactures of Madder." *History and Technology* 12, 1 (1994): 1–22.

Travis, Anthony S. "Decadence, Decline, and Celebration: Raphael Meldola and the Mauve Jubilee of 1906." *History and Technology* 22, 2 (2006): 131–52.

Travis, Anthony S. "From Manchester to Massachusetts via Mulhouse: The Transatlantic Voyage of Aniline Black." *Technology and Culture* 35, 1 (1994): 70–99.

Travis, Anthony S. "Perkin's Mauve: Ancestor of the Organic Chemical Industry." *Technology and Culture* 31, 1 (1990): 51–82.

Travis, Anthony S. "Poisoned Groundwater and Contaminated Soil: The Tribulations and Trial of the First Major Manufacturers of Aniline dyes in Basel." *Environmental History* 2, 3 (1997): 343–65.

Travis, Anthony S. *The Rainbow Makers: Origins of the Synthetic Dyestuffs Industry in Western Europe.* Bethlehem, PA: Lehigh University Press, 1993.

Travis, Anthony S. "Science as Receptor of Technology: Paul Ehrlich and the Synthetic Dyestuffs Industry." *Science in Context* 3, 2 (1989): 383–408.

Travis, Anthony S., and Brent Schools and Industry Project. *The Colour Chemists.* London: Brent Schools and Industry Project, 1983.

Travis, Anthony S., Willem J. Hornix, and Robert Bud. "From Process to Plant: Innovation in the Early Artificial Dye Industry." *British Journal for the History of Science* 25, 1 (1992): 65–90.

Trentmann, Frank. *Empire of Things: How We Became a World of Consumers, from the Fifteenth Century to the Twenty-First.* London: Allen Lane, 2016.

Trentmann, Frank, and Martin Daunton, eds. *Worlds of Political Economy: Knowledge and Power in the Nineteenth and Twentieth Centuries.* Basingstoke: Palgrave Macmillan, 2004.

Trentmann, Frank, and V. Taylor. "From Users to Consumers: Water Politics in Nineteenth-Century London." In *The Making of the Consumer: Knowledge, Power, and Identity in the Modern World*, edited by Frank Trentmann, 53–79. Oxford: Berg Publishers, 2005.

Trevan, J. W. "The Error of Determination of Toxicity." *Proceedings of the Royal Society Biological Sciences* (July 1, 1927): 483.

Tuchman, Arleen Marcia. *Science, Medicine, and the State in Germany: The Case of Baden 1815–1871.* New York: Oxford University Press, 1993.

Turner, R. Steven. "The Great Transition and the Social Patterns of German Science." *Minerva* 25, 1–2 (1987): 56–76.

Turner, R. Steven. "The Growth of Professorial Research in Prussia, 1818 to 1848: Causes and Context." *Historical Studies in the Physical Sciences* 3 (1971): 137–82.

Vaupel, Elisabeth. "Napoleons Kontinentalsperre und ihre Folgen: Hochkonjunktur der Ersatzstoffe." *Chemie in unserer Zeit* 40, 5 (2006): 306–15.

Vaupel, Elisabeth. "Vom Teerfarbstoff zum Insektizid." *Chemie in unserer Zeit* 46, 6 (2012): 388–400.

Vaupel, Elisabeth. "Wissenschaft und Patriotismus: Der Deutsch-Französische Krieg 1870/71." *Chemie in unserer Zeit* 41, 6 (2007): 440–47.

Velkar, Aashish. *Markets and Measurements in Nineteenth-Century Britain.*
Cambridge: Cambridge University Press, 2012.

Von Schwerin, Alexander. "Vom Gift im Essen zu chronischen Umweltgefahren: Lebensmittelzusatztoffe und die risikopolitsche Institutionalisierung der Toxikogenetik in der Bundesrepublik, 1955–1964." *Technikgeschichte* 81, 3 (2014).

Waddington, Keir. *The Bovine Scourge: Meat, Tuberculosis, and Public Health, 1850–1914.* Woodbridge, UK: Boydell Press, 2006.

Waddington, Keir. "The Dangerous Sausage: Diet, Meat, and Disease in Victorian and Edwardian Britain." *Cultural & Social History* 8, 1 (March 2011): 51–71.

Wagner, Tamara S., and Narin Hassan, eds. *Consuming Culture in the Long Nineteenth Century: Narratives of Consumption, 1700–1900.* Lanham, MD: Lexington Books, 2007.

Wainwright, Mark. "Dyes in the Development of Drugs and Pharmaceuticals." *Dyes and Pigments* 76, 3 (2008): 582–89.

Wainwright, Mark. "The Use of Dyes in Modern Biomedicine." *Biotechnic and Histochemistry* 78, 3–4 (January 1, 2003): 147–55.

Walker, Philip D. *Zola.* London: Routledge, 1985.

Waller, R. E. "60 Years of Chemical Carcinogens: Sir Ernest Kennaway in Retirement." *Journal of the Royal Society of Medicine* 87, 2 (1994): 96–97.

Warner, Deborah. *Sweet Stuff: An American History of Sweeteners from Sugar to Sucralose.* Washington: Smithsonian Institution Scholarly Press, 2011.

Webster, Charles. *Victorian Public Health Legacy: A Challenge to the Future.* Birmingham: Public Health Alliance, 1990.

Wehefritz, Valentin. *Bibliography on the History of Chemistry and Chemical Technology. 17th to the 19th Century / Bibliographie zur Geschichte der Chemie und chemischen Technologie. 17. bis 19. Jahrhundert.* Berlin: Walter de Gruyter, 1994.

Wen, Tiffanie. "Food Color Trumps Flavor." *Atlantic,* September 30, 2014.

Wessell, Adele. "Between Alimentary Products and the Art of Cooking: The Industrialisation of Eating at the World Fairs, 1888/1893." In *Consuming Culture in the Long Nineteenth Century, 1700–1900: Narratives of Consumption, 1700–1900,* edited by Tamara S. Wagner and Narin Hassan, 107–23. Lanham, MD: Lexington Books, 2007.

White, Paul. "The Experimental Animal in Victorian Britain." In *Thinking with Animals: New Perspectives on Anthropomorphism,* edited by Lorraine Daston and Gregg Mitman, 59–82. New York: Columbia University Press, 2005.

White, Paul. "Sympathy under the Knife: Experimentation and Emotion in Late Victorian Medicine." In *Medicine, Emotion, and Disease, 1700–1950,* edited by Fay Bound Alberti, 100–124. London: Palgrave Macmillan UK, 2006.

White, Suzanne. "Chemistry and Controversy: Regulating the Use of Chemicals in Foods, 1883–1959." Ph.D. diss.. Emory University 1994.

White Junod, Suzanne. "The Chemogastric Revolution and the Regulation of Food Chemicals." In *Chemical Sciences in the Modern World,* edited by

Seymour H. Maskopf, 322–355. Philadelphia: University of Pennsylvania Press, 1993.

Whorton, James C. *The Arsenic Century: How Victorian Britain Was Poisoned at Home, Work, and Play.* Oxford: Oxford University Press, 2011.

Whorton, James C. *Crusaders for Fitness: The History of American Health Reformers.* Princeton: Princeton University Press, 2014.

Willis, Richard J. B. *The Kellogg Imperative: John Harvey Kellogg's Unique Contribution to Healthful Living.* Grantham: Stanborough Press, 2003.

Wilson, Elizabeth. *Adorned in Dreams: Fashion and Modernity.* New Brunswick, NJ: Rutgers University Press, 2003.

Winstanley, Michael J. *The Shopkeeper's World 1830–1914.* Manchester: Manchester University Press, 1983.

Wisniak, Jaime. "Jean Baptiste Aphonse Chevallier: Science Applied to Public Health and Social Welfare." *Revista CENIC Ciencias Biológicas* 44, 2 (2013): 2–16.

Wood, Donna J. "The Strategic Use of Public Policy: Business Support for the 1906 Food and Drug Act." *Business History Review* 59, 3 (1985): 403–32.

Young, James Harvey. *Pure Food: Securing the Federal Food and Drugs Act of 1906.* Princeton: Princeton University Press, 1989.

Young, R. M. *Darwin's Metaphor.* Cambridge: Cambridge University Press, 1985.

Zeide, Anna. *Canned: The Rise and Fall of Consumer Confidence in the American Food Industry.* Oakland: University of California Press, 2018.

Zylberman, Patrick. "Making Food Safety an Issue: Internationalized Food Politics and French Public Health from the 1870s to the Present." *Medical History* 48, 1 (2004): 1–28.

Index